Everton Diniz dos Santos

PVD-PECVD deposition of silver nanoparticles on textiles

Everton Diniz dos Santos

PVD-PECVD deposition of silver nanoparticles on textiles

Antimicrobial effect and efficiency compared to the Magnetron Sputtering technique

ScienciaScripts

Imprint

Any brand names and product names mentioned in this book are subject to trademark, brand or patent protection and are trademarks or registered trademarks of their respective holders. The use of brand names, product names, common names, trade names, product descriptions etc. even without a particular marking in this work is in no way to be construed to mean that such names may be regarded as unrestricted in respect of trademark and brand protection legislation and could thus be used by anyone.

Cover image: www.ingimage.com

This book is a translation from the original published under ISBN 978-613-9-62810-0.

Publisher:
Sciencia Scripts
is a trademark of
Dodo Books Indian Ocean Ltd. and OmniScriptum S.R.L publishing group

120 High Road, East Finchley, London, N2 9ED, United Kingdom
Str. Armeneasca 28/1, office 1, Chisinau MD-2012, Republic of Moldova, Europe
Printed at: see last page
ISBN: 978-620-8-02118-4

I dedicate this work to my son Caio Cesar Ouverney Diniz dos Santos.

Thank you

I would like to thank my parents, Luiz Carlos Lima dos Santos and Linamar Maria Diniz, for the care, affection and love with which they have always treated me. I would especially like to thank them for their advice on the importance of studying, without which I would certainly not have achieved such a lofty goal. I would like to thank my wife Paula Ouverney Diniz for her support while I was away studying for this course. Thank you to my brother-in-law Leandro Bolzan Rezende, for the enlightening conversations, books and electronics that made my studies and work easier and more feasible. I would like to thank my advisor, Professor Milton Beltrame Jr. for his guidance. I would like to thank Professor Dr Homero Santiago Maciel for his support in scientific writing and in the study of nanotechnology and plasma processes. I would like to thank the rest of my colleagues in the laboratory and Professor Dr$^{\underline{a}}$. Elisa Esposito for her support in microbiology. I would like to thank the other staff and professors at Univap, with whom I had direct or indirect involvement from my undergraduate degree to the end of my master's degree. Finally, I would like to thank the Coordination for the Improvement of Higher Education Personnel (CAPES) for supporting the research carried out during this course.

Knowledge is something that has value in itself, it is the only instrument that truly promotes justice and social transcendence. Everton D. Santos

Summary

Textile materials are used for different purposes by various industrial sectors, especially fashion, construction and architecture. Although the use of these materials is advantageous due to their low market cost and good durability, the contact of textiles with the heat and humidity of the skin makes them a favourable medium for the growth of pathogens. With this in mind, this study aims to compare the characteristics of silver films deposited on cotton textiles using the Physical Vapour Deposition - Plasma Enhanced Chemical Vapour Deposition (PVD-PECVD) technique, usually used to promote the deposition and doping of carbon films, with the information available in the literature on the Magnetron Sputtering (MS) technique. To carry out this experiment, cotton fabrics were coated with silver nanoparticle films (Ag-NPs) produced by PVD-PECVD, and subjected to physical, chemical and biological characterisations. The findings of this study showed points of discrepancy between the properties of the films produced by PVD-PECVD and MS, and also demonstrated a great similarity between the antimicrobial effect produced by both techniques.

Summary

CHAPTER 1

Introduction.

Textile materials are used for different purposes by various industrial segments, especially fashion, construction and architecture. Textiles are usually used for clothing, bed, table and bath, but in hospital environments they are also used to make gowns, masks, coats, hygiene products and injury treatment, among other applications (RADETIC, 2013; CHADEAU et aL, 2010a). Although the use of these materials is advantageous due to their low market cost and good durability, the contact of the fabrics with the heat and humidity of the skin makes them prolific media for the growth of pathogens such as *Staphylococcus aureu* and *Candida a/b/cans* (RADETIC, 2013; EGBUTA; MWANZA; BABALOLA, 2017a; MITRA et aL, 2017a). It is estimated that in the hospital environment, nosocomial infections from textiles are responsible for around 50,000 deaths/year across the European Union, resulting in an average expenditure of US\$ 100 million/year(CHADEAU et aL, 2010a; SZOSTAK-KOTOWA, 2004; RABUZA; SOSTAR-TURK; FIJAN, 2012).

In response to this issue, the textile industry has been incorporating Ag-NPs into fabrics with the aim of making them microbicidal. Various techniques have been used for this, but the Dip-coating process has stood out among the options due to its methodological simplicity, as it only consists of immersing and perpendicularly surfacing the fabric in a solution containing Ag-NPs, and because it requires a low investment in equipment. However, this process has some disadvantages, such as using polluting volatile solvents, Ag-NPs supplied by third parties and therefore susceptible to loss and/or variations in quality, and being sensitive to small vibrations in the environment that compromise the homogeneity of the films (DAVID et aL, 2017; ZHANG et aL, 2013; TANG et aL, 2017; RADETIC, 2013).

When compared to plasma film deposition technologies such as the MS process, the Dip-coating technique is at a clear disadvantage, since MS processes have all the positive qualities of the Dip-coating process, and are also able to simultaneously produce and fix Ag-NPs on different substrates, have a high deposition rate, reduced production time and no waste resulting from the process (ZHANG et al, 2013; KELLY; ARNELL, 2000a). The only advantage that the dip-coating process has over the MS technique is the control of the film's characteristics. This problem has been overcome by hybridising the PECVD technique, which offers extensive control over film variables, with the MS technique, resulting in a process known in the literature as ASP- PECVD, PVD-PECVD or even PVD/PECVD(R et aL, 2015; DURST; ELLERMEIER; BERGER, 2008).

In general, the PVD-PECVD technique has only been used to promote doping and deposition of carbon coatings - carbon nanotubes and Diamond-like carbon (DLC) - which are more uniform than those obtained industrially by carburisation, nitriding and carbonitriding techniques (MINEA et al., 2005; ZEMLIcKA et aL, 2016; VENKATESH; TAKTAK; MELETIS, 2014). As presented above, this technique has characteristics that are promising for the deposition of Ag-NPs in tissues, however no studies have been carried out in this regard to date.

The aim of this work is to produce Ag-NPs coatings on cotton fabrics using the PVD-PECVD technique, to determine their physical, chemical and biological properties, and to compare them with the properties usually found in the literature for films produced by MS. To this end, the logical sequence of the experiments in this work was divided into three sections: experimental, which presents the methodological procedures used to conduct the experiments; results and discussion, which analyses and discusses the data obtained from the experiments; and finally, conclusions, limitations and suggestions for further research.

This study was made up of three major blocks: sample production, film characterisation and comparison between the techniques (Figure 1). These groups together comprise seven stages of the experiment, which produced data that was subsequently presented, analysed and discussed in Chapter 4. In this way, each stage of the experiment and the discussion contribute to answering the questions raised. The aspects of the procedures and the parameters used in each of the stages presented in Diagram 1 are described in more detail in Chapter 3. The conclusions drawn from the results are presented in Chapter 4.

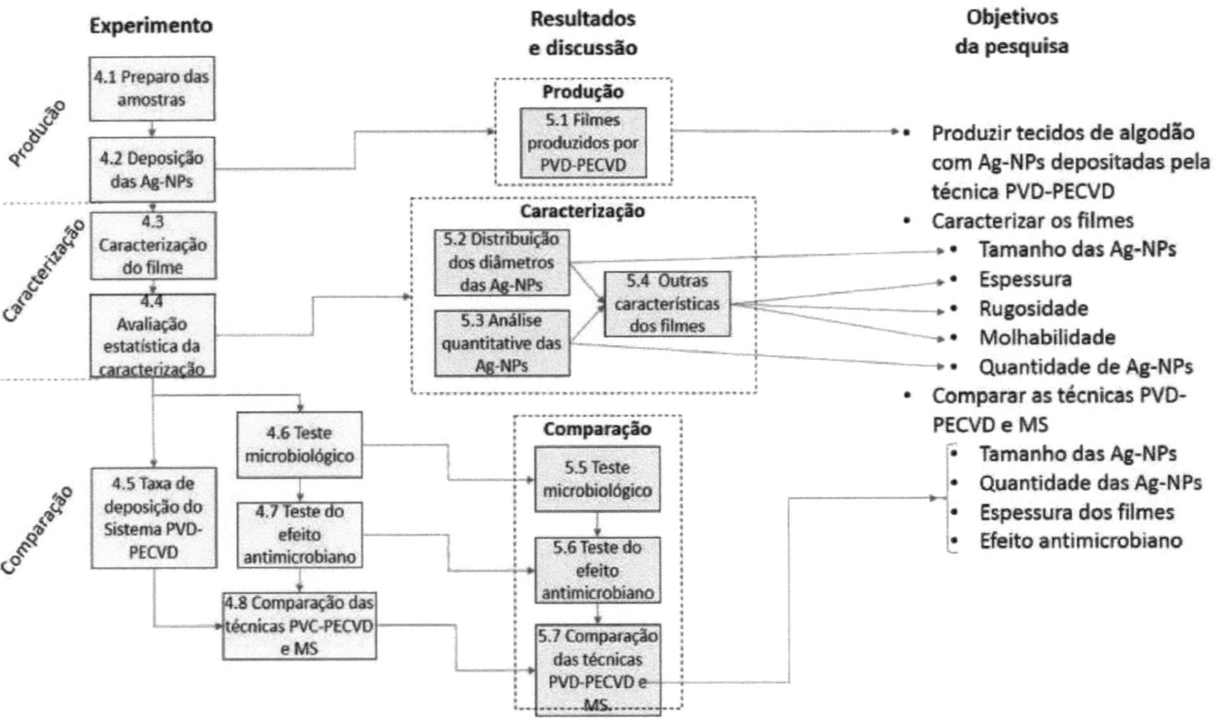

Figure 1-0 The diagram shown in this figure represents the entire logical chain of this study. In this way it is possible to understand exactly how each stage of the experiment relates to the research objectives and contributes to the discussion of the results.

8

CHAPTER 2

Objective.

2.1   General Objective.

To determine the relative efficiency of the PVD-PECVD technique in the field of Ag-NPs film production on textiles.

2.2   Specific Objective.

- To check the feasibility of producing Ag-NPs coatings on cotton fabrics using the PVD-PECVD technique; - To check the feasibility of producing Ag-NPs coatings on cotton fabrics using the PVD-PECVD technique.

- Determine the physical, chemical and antimicrobial properties of the films produced by PVD-PECVD;

- Compare the results produced by the PVD-PECVD technique with the results available in the literature for the MS technique.

CHAPTER 3

Literature review.

This chapter presents the state of the art, covering the definitions of the materials and techniques used in this experiment. Subsection 3.1 presents a historical survey of the use of cotton garments, while subsection 3.2 discusses the characteristics of the plant that produces cotton, the cotton crop and its stages prior to commercialisation. Subsection 3.3 presents a survey of the main microorganisms carried by cotton fabrics, especially in hospital environments. Subsection 3.4 discusses silver nanoparticles, shedding light on their history, physical properties and biological effects. Subsection 3.5 presents the history and mechanisms of the dip coating process in the textile industry. Subsection 3.6 closes the literature review by discussing plasma silver film deposition techniques, with a special focus on MS, PECVD and PVD-PECVD techniques.

3.1 First cotton garments.

Anthropological studies carried out on human fossils dating back more than 1 million years suggest that the bodies of these individuals were completely covered in hair that protected them against both daily and seasonal temperature variations (Figure 2) (L'ABBÉ et al., 2015).

Figure 2 - Representative engraving of a typical Australopithecus family.

http://www.ancient-origins.net/news-evolution-human-origins/evolution-human-birth-incredible-story-mil lion-years-making-007976

As the hairs that cover contemporary animals are linked to barium receptors, it's quite reasonable to assume that in the distant past they were also linked to barium receptors.

their function was to help perceive imminent dangers, such as the proximity of small venomous insects (JUNQUEIRA; CARNEIRO, 2013). However, to the detriment of these advantages, the history of human evolution suggests the occurrence of several significant modifications to the body structure of the first hominids, of which, in the context of this dissertation, the loss of hair during the

10

millennia of transition from Australopithecus to Homo sapiens stands out (Figure 3)(PFENNIG; PFENNIG, 2010; CAPDAREST-AREST; GONZALEZ; TURKER, 2014; STRINGHAM et aL, 2012). Without fur, climate change became a major problem for hominids, who therefore began to use animal skin as clothing (FABRICATING...,2011; ZHANG et al., 2016).

Figure 3 - Human evolution, from a common ancestor with the ape to *Homo sapiens.*
http://guiacemtiradentes.blogspot.com.br/2013/08/evolucao-biologia.html

Over time, human settlements grew into large cities, the first of which, according to archaeology, were located in the valley between the Tigris and Euphrates rivers in North Africa, and also along the course of the Nile river in Egypt. Cities such as Ur, Uruk and Thebes became economic, political and cultural exponents of the people of antiquity, who did not appreciate wearing clothes made from animals, both for religious reasons and because of the discovery of cotton and its textile properties (GROF et al, 2013; OLIVEIRA et al, 2007; REDFORD, 1993; SHAW, 2002).

Today cotton fabrics are, among other applications, widely used throughout the world for clothing, bedding, tableware and bath products, which are responsible for fuelling an industry with an annual turnover of more than R$4.7 billion (CHADEAU et aL, 2010b; RADETIC, 2013; LEITE et aL, 2015; LIMA et aL, 2006; CHAN; OLIVEIRA, 2010).

3.   2Cotton.

Cotton is a white fibre of plant origin generated around the seeds of the *Gossypium sp.* plant. This fibre is whitish, soft, grows and protects the seeds of the plant in a similar way to the fruit produced by angiosperms, such as apples, guavas, oranges and others (Figure 4)(SOARES; MACIEL, 2001; AVCI et al., 2013).

Figure 4 - Representative image of the Gossypium species with cotton buds. https://american-seed.com/cotton-varieties-eyed-agrilife-extension-trials-across-texas-high-plains/

The *Gossypium sp.* plant has a tropism for warmer, rainier regions of the planet. Consequently, the closer a country is to a tropical climate, the more favourable it is to grow cotton in its territory. Brazil currently occupies fifth place in the ranking of cotton producing countries, with an annual production of 1,639,537 tonnes. This ranking is led by China, with a production of 6,841,593 metric tonnes per year, and India with a production of 5,323,467 metric tonnes annually (ECHER et aL, 2008; ALVES et aL, 2006; PASINATO et al., 2016).

After correctly fertilising the soil and correcting its pH, which are essential steps for any crop, cotton seeds are planted in the soil and irrigated daily. The process of plant growth and the availability of cotton buds for harvesting takes place between 140 and 150 days after planting, and can be done manually or mechanised. Certainly, each method has its inherent advantages and disadvantages, for example: manual harvesting is costly, but the quality of the cotton selected is very high.

Mechanised harvesting is less expensive and quicker than manual harvesting, but it produces less pure products that can withstand shorter storage times and require more work to process. Depending on the level of purity of the harvested product, the suitability of the pieces for commercialisation as fabric will depend on the execution of 8 stages (PASINATO et al., 2016; ALVES et al., 2006):

1) Ginning: after harvesting, the cotton buds collected have seeds inside, especially those from the mechanised harvesting process. These seeds are removed manually during this stage.

2) Spinning: With the cotton free of impurities and seeds, the fibres are braided with different thicknesses to form threads.

3) Weaving: The yarns produced in stage 2 are woven on looms to form fabrics.

4) Flaming: The fabric pieces are passed through flames to remove excess fibres. The purpose of this stage is to improve the appearance and texture of the garments.

5) Bleaching: the fabrics are bathed in bleach to lighten them. This stage is especially interesting in cases where the fibres have a yellowish colour or vary greatly in tone.

6) Mercerisation: cold application of caustic soda which reacts with the cellulose of the cotton and increases the strength, shine, durability and flexibility of the fabric.

7) Dyeing: changing the colour of fabric by treating it with dye.

8) Finishing: the fabric goes through various chemical products to gain resistance and protection against harmful agents.

The main characteristics of cotton fabrics are: easy handling, high comfort, quick drying and a high capacity to absorb liquids. This material also has a low tendency to cause allergic reactions, is versatile as it can be mixed with other fibres to form a material with new properties, and is highly resistant to mechanical cleaning processes. Cotton fabrics are ideal for hot climates, as their moisture absorption capacity is 8 per cent. In terms of use, cotton fabrics are used for clothing, tapestries, sewing threads, bed linen, table linen and bath linen, among others (RADETIC, 2013; CHADEAU et al, 2010b).

To the detriment of all the advantages and benefits of using cotton to make fabrics, this fibre has also proved to be a very prolific medium for the development of microorganisms, especially in conditions of high humidity, which are very common in clothing and bed, table and bath linen (RADETIC, 2013). The next sub-section will address this issue in greater depth, listing the main microorganisms present in cotton textiles exposed to high humidity.

1.1 Nosocomial infections.

In most situations where cotton fabrics are used for covering or clothing, one of the advantages of this material is its high capacity to absorb liquids. However, it is not always feasible to sanitise the fabric immediately after use, either directly, such as cleaning a damp surface, or indirectly, such as a garment that absorbs sweat on a sunny summer's day. As the chemical composition of cotton fibres contains a wide variety of organic compounds, prolonged exposure of a cotton textile to moisture encourages the proliferation of microorganisms around the fibres of this material (RADETIC, 2013; CHADEAU et al., 2010b) (RADETIC, 2013; CHADEAU et al., 2010b). This situation is especially harmful in a hospital environment, since the transfer of pathogens from one hospital sector to another, or even to the outside environment, provides a source of disease dissemination (MITCHELL; SPENCER; EDMISTON, 2015; LÓPEZ-GIGOSOS et al., 2014). Among the specimens of microorganisms carried by cotton textiles, Pseudomonas aeruginosa (PANAGEA et al., 2005), Enterococcus faecium (CHENG et al., 1997; KUZDAN et al., 2014), and methicillin-resistant Staphylococcus aureus (MRSA) (CRUZ et al., 2008; MIYAZAKI; BÔAS, 2006) stand out in medical reports due to their associated biological risk.

With regard to P. aeruginosa, its transmission to patients occurs through numerous routes, including direct contact through ingestion of contaminated water, inhalation of aerosols, contact with contaminated medical devices (instruments, materials and tissues), and through the hands of healthcare professionals when poorly sanitised (JONES; LUTZ, 2014; CLIFTON; PECKHAM, 2010). The spores of this microorganism resist very low humidity conditions for up to six months (KRA-MER; SCHWEBKE; KAMPF, 2006). Another microorganism also transmitted by cotton textiles are strains of Enterococcus faecium. This bacterium forms short-chain Gram-positive cocci exhibiting alpha haemolysis or the absence of haemolysis. This bacterium is a commensal organism of the human intestine, and is an important causative agent of healthcare associated infections. This microorganism is responsible for many cases of medical complications, especially when the patient is contaminated with vancomycin-resistant strains (VRE) (KUZDAN et al, 2014; CHENG et al, 1997). E. faecium is a microorganism that is tolerant to a wide range of environmental conditions, being able to withstand heat treatment or washing for up to 60 minutes, a wide variation in pH, and high concentrations of sodium (6.5%) and bile salts (FIJAN et aL, 2007; LAPORT et aL, 2003). On the other hand, strains of this microorganism are used as pro-biotics in humans and animals (SISSON et aL, 2014; FIJAN, 2014) and are even present in the microflora of many varieties of cheese (SABIA et aL, 2008; AL-MARIRI; YOUNES; SHARABI, 2013). Finally, with regard to Staphylococcus aureus specimens, Gram-positive and coagulase-positive commensal cocci capable of colonising the nose and skin, their transmission to immunologically weakened individuals can lead to serious illnesses such as septicaemia, pneumonia and meningitis (JENSEN et aL, 2015; MATUSSEK; TAIPALENSUU; EINEMO, 2007). This microorganism is especially harmful when the strains are of the methicillin-resistant type (MRSA), as this strain is more selective in terms of the antibiotics that are effective in neutralising it. S. aureus spores normally survive up to 7 months in low humidity conditions.

The most effective way of preventing the transfer of pathogens from textiles to other patients is immediate chemo-thermal disinfection, or thermal washing procedures with water and disinfectant agents (FIJAN; SOSTAR-TURK; CENCIC, 2005; FIJAN et al, 2007; ALTENBAHER; TURK; FIJAN, 2011). However, it is not always possible to wash all hospital materials and instruments on a daily basis, which is why the use of treated textile fibres and surfaces treated with antimicrobial substances, such as Ag-NPs, is the most effective and feasible prophylactic measure (SHARMA; YNGARD; LIN, 2009; SAMBERG; ORNDORFF; MONTEIRO-RIVIERE, 2011; EREMENKO et aL, 2016).

The next subsection will go into more detail about the characteristics and properties of Ag-NPs.

1.2 Silver nanoparticles.

Ag-NPs confer silver particles with a diameter measured on the nano-metre scale, which, as a result of their simple dimensions, have potentially interesting characteristics for fighting microorganisms (EREMENKO et al, 2016; ZILLE et al, 2015). However, even bodies made up of a large mass of Ag, such as tools, household utensils and objects of adornment, have remarkable antimicrobial effects. More than two millennia ago, during the height of classical antiquity, the Greeks and Romans already used silver as an amulet to protect their provisions because, for example, the continuous contact of a piece of silver with milk significantly slowed down its decomposition process. Naturally, the people of antiquity didn't know the cause of this effect and attributed mystical powers to this metal (SOUZA et al, 2013).

It is currently estimated that Ag-NPs are effective against around 650 pathogenic microorganisms, including strains resistant to antibiotics commonly available on the market (DASTJERDI; MONTAZER, 2010). In addition to the antimicrobial properties of silver, Ag-NPs have optical and catalytic properties that enable this material to be used to manufacture sensors, energy converters, electronic chips, among others (MORAES et aL, 2015; BUDAMA et aL, 2013; HOMAN et aL, 2010).

Due to the versatility, low cost and promising characteristics of Ag-NPs in the medical field, extensive research has been carried out on this metal in recent decades (BAKI et al, 2014; HOLBROOK; RYKACZEWSKI; STAYMATES, 2014; CHIDAMBARAM; KRISHNASAMY, 2014).As Figure 5 shows, the growth profile of articles and citations on the topic of "silver nanoparticles" has undergone a progressive annual increase throughout its historical series, delimited at the bottom by the year 1991 and at the top by the current year 2018. According to the records made available by Thomson Reuters, the agency that manages the Web Of Science, some 58,742 papers relating to the study of silver nanoparticles have been published in the last 27 years. Of this total, 4,700 studied the antimicrobial effects of Ag-NPs, 441 of which correspond to studies of Ag-NPs in textiles (THOMSON REUTERS, 2018).

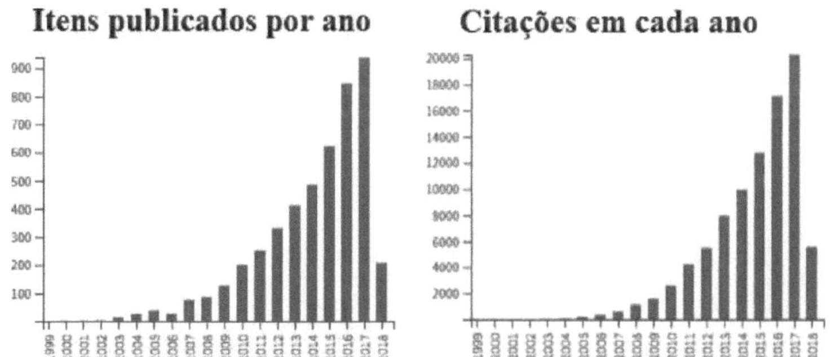

Figure 5 - Growth profile of research and citations on the subject of "silver nanoparticles".
https://apps.webofknowledge.com/CitationReport.do?product=WOS&search_mode=CitationReport&SID=5D6SoxMnG6v1sLZThJT&page=1&cr_pqid=2&viewType=summary&colName=WOS

However, even with all this research, the scientific community has yet to establish a consensus on the most effective technique for producing and/or fixing Ag-NPs on textiles, since the process most effectively used in industry, the dip coating process, ends up increasing the relative cost of garments, making it difficult for the public sector in particular to take advantage of this technology. Therefore, the search for a more efficient process than the current one favours making this technology available to all sections of society. In this context, plasma technologies, such as PECVD and Magnetron Sputtering (MS) systems, have gained great prominence, as described in subsection 3.5, on the dip coating technique, and subsection 3.6, on electric plasma deposition processes (DAVID et aL, 2017; ZHANG et aL, 2013; TANG et aL, 2017; RADETIC, 2013).

1.3 Dip-coating process.

The dip coating process is the most widely used by industries looking to incorporate Ag-NPs into textiles. The production of films using this technique takes place in three stages: first the substrate is immersed vertically in a liquid system of nanoparticles or other elements in solution, and then it is removed at a well-defined constant speed; once this stage is complete, the excess solvent is removed by evaporation and/or draining. The dip coating technique is used both in batches of substrates and in continuous production. According to Brinker, Clark and Ballance (1988), this methodology can be divided into 5 stages in batch production (Figure 6 a,b,c,d,e), and one stage in continuous production mode (Figure 6 f): (a) immersion of the artefact in the bath; (b) start of removal of the artefact from the bath; (c) deposition and draining of the coating on the artefact; (d) gravitational flow of the excess coating and (e) evaporation of the solvent from the bath. With regard to the continuous process, shown in Figure 6 f, its degree of complexity is reduced compared to batch production, since the formation of the coating occurs constantly and, in a way, masks the drainage stage in the deposition of the film.

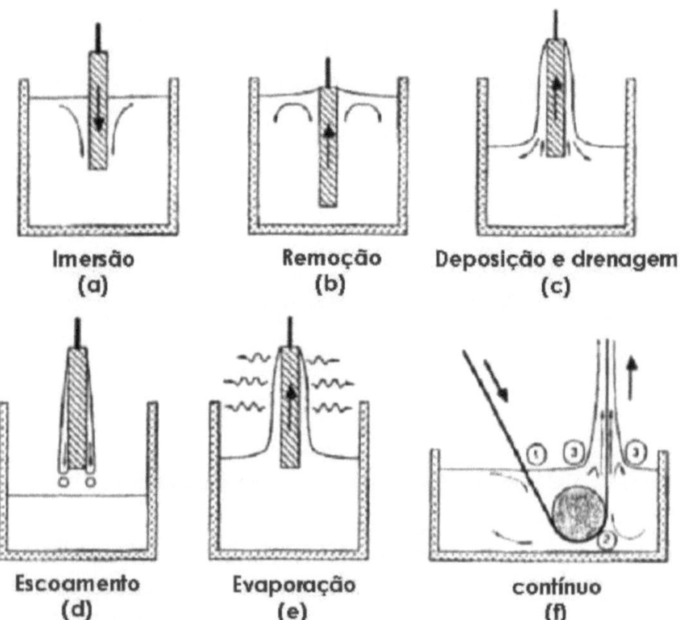

Figure 6 - Stages of the dip coating process in batches (a, b, c, d, e), and continuous (f). (Adapted from Brinker, Clark and

Ballance (1988)).

The formation of thin films by dip coating represents the oldest commercial application for sol-gel technology. The first patent based on this process belongs to Jenaer Glaswerk & Gen. in 1939 for the production of silica films (C.J.BRINKER et al, 1991). The first records of the industrial use of this technique date back to the production of candles, since this material was developed by dipping a piece of string in liquid paraffin successively until it reached the desired calibre. Other industrial sectors have also used this technique to produce films on ceramic artefacts, paint surfaces, coat tools with rubber, and finally, to incorporate Ag-NPs into fabrics (DAVID et al., 2017; ZHANG et al., 2013; SILEIKAITE et al., 2006). The dip-coating process is still the textile industry's most popular method for incorporating Ag-NPs into fabrics, but some companies have replaced it with more environmentally friendly and efficient processes based on electric plasmas, as described in more detail in the next subsection.

3.6 Plasma production and deposition techniques   for Ag-NPs.

Various methodologies for different purposes use electric plasmas as their basic operating principle. Plasmas can be understood as highly energetic fluids made up of a mass of free electrons, cations and anions (Figure 7)(ASTAPENKO, 2013; MOURA, 2007; FERRARI; ROBERTSON, 2000).

Figure 7 - Photograph of a plasma generated between two electrodes.

http://physicsworld.com/cws/article/news/2003/apr/30/cold-plasmas-destroy-bacteria

However, with regard to the deposition of Ag-NPs on textiles, the methodologies that stand out the most are MS and PECVD. Subsection 3.6.1 will deal with Sputtering and Magnetron Sputerring, while subsection 3.6.2 will deal with the PECVD technique.

3.6.   1Sputtering and Magnetron Sputtering.

In order to properly understand the MS film deposition methodology, it is first necessary to understand how the sputtering technique works, since the MS technique is just a variation of this methodology. The sputtering deposition process produces films by bombarding ions from the plasma produced by a voltage source onto an electrode that can be made of different metals. In the case of an experiment to produce nanostructured silver films using this technique, the plasma ions would collide with a silver electrode. As a result of this collision, countless nanometre-sized particles would be ejected from the electrode towards the substrate, thus producing Ag-NPs. These Ag-NPs are deposited in overlapping layers to form the film. The deposition rate of Ag-NPs using this technique is quite low, however

can be improved by positioning magnets behind the targets and thus increasing the mean free path and the electron spiral. These modifications result in an increase in the energy of the collisions, consequently significantly increasing the quantity of particles ejected from the electrode with each collision. When the sputtering system is modified by the presence of strong electromagnets, it becomes known as magnetron sputtering (Figure 8) (KELLY; ARNELL, 2000b; ROBERTSON, 2002).

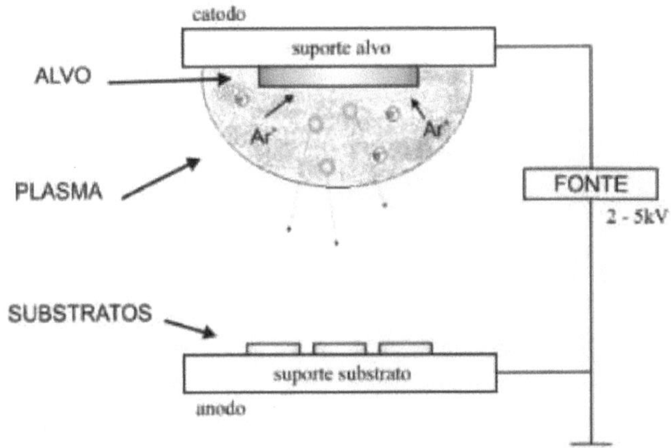

Figure 8 - Sputtering deposition scheme.
http://www.univasf.edu.br/~joseamerico.moura/index_arquivos/Cap3.pdf

The MS technique has important advantages over the dip-coating technique, as it does not require the use of chemical solvents, making it safer and more environmentally friendly, and has a high silver deposition rate. This technique is not sensitive to disturbances in the environment as the dip coating technique is. The MS technique is also capable of producing Ag-NPs, whereas the dip coating process does not. However, although the MS technique has several advantages over the dip coating method, the time it takes to establish the right vacuum for the working conditions in this system and its low versatility have encouraged research into the PECVD system, which seems to meet these demands (DOWLING et al., 2001; SILEIKAITE et al., 2006; RIVERO et al., 2011).

3.6.2PECVD .

Better known as "Plasma Enhanced Chemical Vapour Deposition (PECVD)", PECVD processes are understood as hybrid techniques resulting from the relationship between physical

(PVD) and chemical (CVD) deposition techniques. In this way, characteristics that are notoriously common to the original processes are also found in the PECVD technique. Like all PVD, PECVD works at low pressures, around 10^{-2} Torr, under the action of high potential gradients. Whereas CVD plasmas have a high temperature. In PECVD, these types of plasmas are also generated, but by a large potential difference at a low pressure, which causes the internal energy of the electrons to rise to the order of 100-300 eV, changing the gaseous aggregation phase to the plasma state (Figure 9)(MARCIANO L.F. BONETTI, 2009; ROBERTSON, 2002; CASIRAGHI; FERRARI; ROBERTSON, 2005).

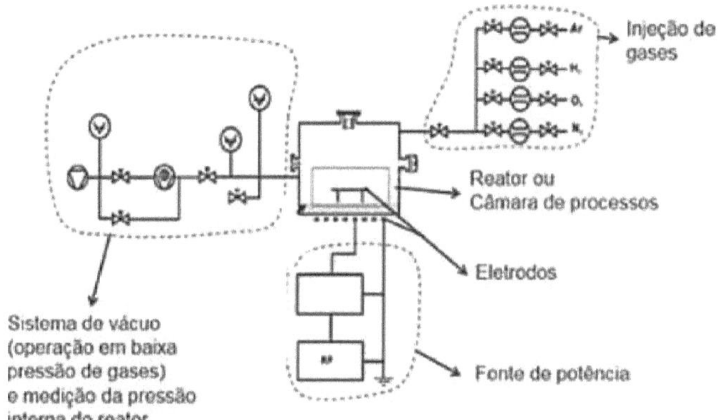

Figure 9 - Schematic of a PECVD reactor with a radio frequency (RF) source.
Adapted: Prof Dr Rodrigo Sávio Pessoa

Generally speaking, the formation of plasma in both this and other systems begins with a small quantity of free electrons that are present in the gas due to external radiation. With the application of an external electric field, generated by the cathode and anode, the free electrons acquire sufficient energy and collide with the neutral particles in the gas, promoting the generation of new electrons through ionising collisions. At the same time, electrons are lost to form negative ions. When the number of electrons is sufficient to produce only the ions needed to regenerate the number of electrons lost, a stable state is reached (the luminescent state of the plasma is generated) in which a balance is established between the rate of ion formation and the rate of recombination of ions with electrons. Thus, the generation and sustaining of a plasma, as well as its chemical kinetics, are the result of collisional processes involving neutral and charged particles, as well as diffusion losses (PESSOA, 2009). The collisions that occur between particles in the plasma are divided into two types (LIEBERMAN, 2005): 1) elastic collision: where there is only a transfer of momentum between the particle

2) inelastic collision: where there is an exchange of energy between the incident particle and the target. The energies involved in this exchange can vary from less than 0.1 eV (e.g. vibrational excitation) to more than 10 eV (e.g. ionisation). Table 1 lists the main reactions involving collisional processes between heavy particles and electrons, and between ions and neutral atoms, respectively

(PESSOA, 2009).

Table 1 - Reactions occurring in the gas phase involving neutral/ionic species and electrons. In this table, atoms and radicals are denoted as A and B; molecules as AB; excited species, at energy levels above the fundamental level, with the superscript *; and positive and negative ions with the superscripts "+" and respectively.

Reactions	Types of process
1 $e^- + A \rightarrow A + e$	elastic spreading
2 $e^- + A \rightarrow A^+ + 2e^-$	Ionisation
3 $e^- + A \rightarrow A^* + 2e^-$	excitation (formation of metastables)
4 $e^- + A^* \rightarrow e^- + A + h\nu$	de-excitation
5 $e^- + A^* \rightarrow A^+ + 2e^-$	two-step ionisation
6 $e^- + AB \rightarrow A + B + e^-$	dissociation
7 $e^- + AB \rightarrow A^+ + e^- + B + e^-$	dissociative ionisation
8 $e^- + AB \rightarrow A + B$	electron capture with dissociation
9 $e^- + A^+ + B \rightarrow A + B$	recombination in volume
10 $A^+ + B \rightarrow B^+ + A$	(resonant for B = A)charge transfer
11 $A^+ + B \rightarrow B + A^+$	elastic spreading
12 $A+ + B \rightarrow A^+ + B^* + e^-$	excitement
13 $A^+ + B \rightarrow A^+ + B^+ + e^-$	ionisation
14 $A + B^* \rightarrow A^+ + B + e^-$	Penning ionisation
15 $A^+ + BC \rightarrow A^+ + B + C$	dissociation
16 $A^{+/-} + B \rightarrow AB^{+/-}$	oligomerisation
17 $A + B \rightarrow AB$	oligomerisation
18 $e + A^{-+} + B \rightarrow A + B$	recombination in volume

Films produced using the PECVD technique are highly dependent on the adjustment of numerous parameters used in the process, such as: voltage, current, gas flow, chamber pressure, process duration, cathode temperature and plasma power (ROBERTSON, 2002). As in the PECVD process the substrates are positioned over the cathode, they are completely surrounded by the plasma sheath region. Because of this, the distribution of the films over the substrates occurs with significant homogeneity (HETTLICH et al, 1991; WILD et al, 1994; MARCIANO et al, 2009).

When compared to plasma film deposition technologies, the dip coating technique is at a clear disadvantage because MS processes have all the positive qualities of the dip coating process, and are also able to simultaneously produce and fix Ag-NPs on different substrates, have a high deposition rate, reduced production time and no waste resulting from the process (ZHANG et al, 2013; KELLY; ARNELL, 2000a). The only advantage that the dip coating process has over the MS technique is the control of the film's characteristics. This problem has been overcome by hybridising the PECVD technique, which offers extensive control over film variables, with the MS technique, resulting in a process known in the literature as ASP-PECVD, PVD-PECVD or even PVD/PECVD (R et aL, 2015; DURST; ELLERMEIER; BERGER, 2008).

In general, the PVD-PECVD technique has only been used to promote doping and deposition of carbon coatings - carbon nanotubes and Diamond-like carbon (DLC) - which are more uniform than those obtained industrially by carburisation, nitriding and carbonitriding techniques (MINEA et al, 2005; ZEMLIcKA et al, 2016; VENKATESH; TAKTAK; MELETIS, 2014). As presented above, this technique has characteristics that are promising for the deposition of Ag-NPs in tissues, however no studies have been carried out in this regard to date.

CHAPTER 4

Methodology.

This chapter presents the experimental procedures and methodologies used to carry out this study. Subsection 4.1 presents the pre-treatments received by the substrate in order to make them suitable for receiving the silver coating described in detail in subsection 4.2. The films were characterised with regard to their physical and chemical properties in accordance with the procedures detailed in subsection 4.3. The statistical processing of the characterisation results is detailed in subsection 4.4. Subsection 4.5 details the procedures adopted to determine the deposition rate of the PVD-PECVD system. The microbiological tests carried out on the samples are detailed in subsection 4.6, while the tests to determine the type of antimicrobial effect are presented in subsection 4.7. In the last subsection of this chapter, subsection 4.8, the MS and PVD-PECVD techniques were compared in terms of their similarities and differences.

4.1   Sample preparation.

The experiment began with the preparation of a set of 80 samples of unbleached cotton fabrics measuring 1.5 cm^2 and 1 mm thick, which were distributed equally among 16 groups. Each group consisted of 5 elements which were sent for plasma treatment. The sample set underwent a mechanical cleaning process with neutral detergent and water, and then an acetone bath for 10 minutes in a QUIMIS Q335D ultrasound system. The cleaning procedures were finalised by drying the samples for one hour at a temperature of 60° C in a Sterilifer model SX 1.0 DTME digital oven. Once this stage was completed, the samples were ready to receive the silver film through the deposition process.

4.2   Silver deposition process.

The previously prepared samples were subjected to the silver film deposition process using the PVD-PECVD technique. All the procedures were carried out in a pulsed DC source PECVD system manufactured by NA-NOMASTER, model NPE-4000. This system was hybridised with MS by positioning a hollow cathode, called Active Screen Plasma (ASP) (R et aL, 2015), on the cathode of the PECVD system. The deposition process was subdivided into three stages: cleaning the ASP (Figure 10-1), positioning and fixing the substrates and the ASP (Figure 10-11), and plasma rupture (Figure 10-111). With regard to stage I, the ASP was subjected to a mechanical cleaning process using 300 grit water sandpaper. At the end of this stage, the piece was washed with neutral detergent and running water, and dried for one hour at a temperature of 100°C in a Sterilifer model SX 1.0 DTME digital oven. In step II, the samples were positioned between a ceramic disc and a glass slide and then placed on the PECVD cathode, while the ASP was positioned over the samples to keep them in the centre. As for stage III, in order to remove reactive gas species and water droplets that would jeopardise the reaction, the internal pressure of the PECVD system was reduced to 10^6 torr by activating the turbo (Analger model 800) and scroll (Edwards model E2M18) vacuum pumps simultaneously. Subsequently, the pressure was raised to 3×10^2 torr by injecting oxygen gas to sterilise the entire system for 10 minutes (LUQUETA et al., 2016). At the end of this period, the oxygen was replaced by argon to start the Ag-NPs deposition process (Figure 10). Table 2 shows the general parameters maintained during this process, of which only the deposition time and plasma power were varied depending on the group to be produced. The gases used in the process are 99% pure and were supplied by White Martin.

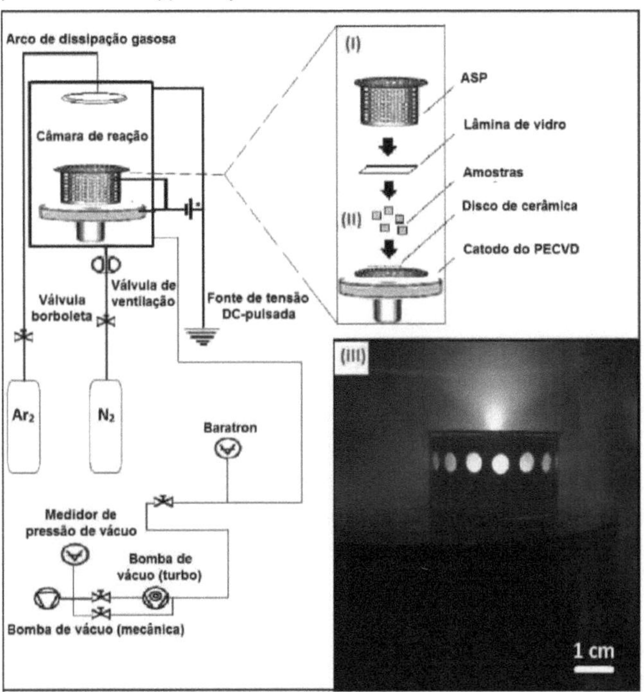

Figure 10 - (I) positioning the ASP on the samples, (II) positioning and fixing the substrates on the PECVD cathode, (III) plasma rupture; the gases entered the system through the dissipation arc; the internal pressure was measured by the baratron sensor; the voltage source (DC-pulsed) was activated to generate the plasma; nitrogen gas was used to re-establish the balance between the internal and external pressure of the chamber for its opening; the flow of gases was adjusted by butterfly valves.

Table 2 - Parameters kept constant during the silver deposition process.

Parameters	Value
Pressure (Torr)	3×10^{-2}
Pulse (KHz)	100
Reverse pulse (us)	0,8
Voltage (V)	650
Current (A)	0,6
Temperature variation (°C)	18 a 50
Gas	Argon

Once the film production stage was completed, the coated samples were characterised using various techniques, as outlined in the following subsection.

4.3 Film characterisation.

The silver film characterisation stage began with the morphological analysis of the Ag-NPs-coated tissue fibres, using high-resolution images produced by scanning electron microscopy (SEM) on an FEI Inspect F50 system. Images were produced using backscattered electrons and secondary electron emission, respectively, with an accelerating voltage of 20 kV and 3 kV. The Ag-NPs were quantified by the percentage by weight of silver atoms (PPAg) present in the total chemical composition of the films. To do this, the samples were previously coated with a conductive Au-Pd film (80-20 % by mass), and then a total of 80 spectra were taken from quintuplicates of the sample groups, obtained by energy dispersive X-ray spectroscopy (EDX). The EDX detector used in this research is the same make and model as the SEM (Inspect F50), as it is a module attached to this equipment. The PPAg was determined by the arithmetic mean of the values found in the quintuplicates of each group.

The quantity and diameter of the Ag-NPs were determined by computer analyses of images of the samples produced by SEM. To read this data, the images were processed in the IMAGE J software in three stages: (I) colour equalisation and application of filters, (II) data segmentation, and (III) counting the number and size distribution of Ag-NPs. During stage (I), the Ag-NPs were emphasised to the detriment of the cotton substrate. The pixels corresponding to the Ag-NPs were marked during step (II) to be counted during step (III), using the "particle analyses" and "distribution analyses" commands (Figure 11).

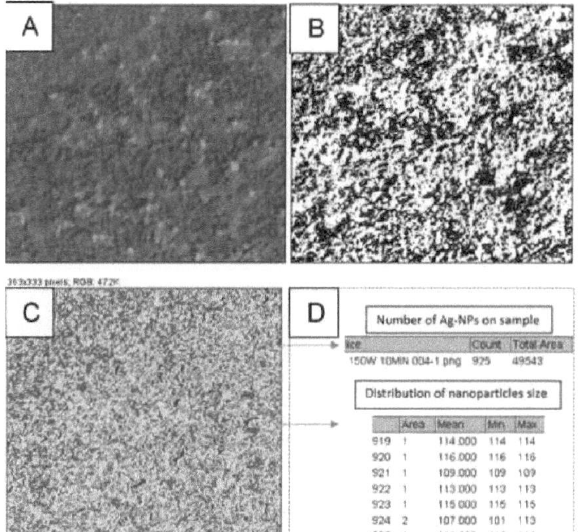

	Area	Mean	Min	Max
919	1	114.000	114	114
920	1	116.000	116	116
921	1	109.000	109	109
922	1	113.000	113	113
923	1	115.000	115	115
924	2	107.000	101	113
925	1	116.000	116	116

Figure 11 - Image processing in the IMAGE J software: (A) image in its original state, (B) image treated by subtracting the background and highlighting the Ag-NPs in black, (C) data segmentation with the Ag-NPs highlighted in yellow, (D) table representing the results from the analyses in stage III.

The thickness of the films and the maximum amplitude of the roughness of the samples were determined by 3D optical profilometry on a KLA-Tencor system, capturing data using 21 and 11 traces of 199,797 [im, respectively in the groups covered and not covered by Ag-NPs, at a pass rate of 100 um/s and a frequency of 200 Hz. The thickness of the films was directly measured by the equipment and the maximum amplitude of the roughness of the samples was determined by calculating the difference between the roughness of the group with the highest PRAg and the roughness of the tissues not covered by the film. As a result, measurements were made on 10 samples, corresponding to quintuplicates of each group, and expressed as the mean ± standard deviation of the root mean square (RMS) of each group, which expresses the statistical measure of the magnitude of the roughness variation along the surface of the samples.

The wettability of the surface of the samples was determined based on the principle of the sessile droplet, by analysing the static contact angle (CA) of the water droplet (polar liquid) (ZILLE et al., 2015). The volume of the drops was kept constant at 2 [iL. All quintuplicates of the samples were analysed on the Kruss DSA 100 equipment between 3 and 14 days after the coatings were produced. The data collected was reported as a function of the arithmetic means ± standard deviation determined.

Having completed the film characterisation stage, each group of values was subjected to statistical tests as described in detail in the next subsection.

4.4 Statistical analysis of characterisations.

The values reported in the sample characterisation stage were all organised according to their arithmetic means ± standard deviations of the values measured in the quintuplicates of each test carried out on the samples: morphological analysis of the fabric fibres covered with Ag-NPs, determination of PPAg in the chemical composition of the films, quantity and diameter of the Ag-NPs, film thickness, maximum amplitude of the roughness of the samples, and surface wettability. This measure of central tendency and variance was chosen to compile the data resulting from the 80 samples analysed in each test into just 16 representative samples for each characterisation.

With regard to the correlation between the time/power variables and the PPAg in the films, the distribution of the data relating to the silver content in the samples was normalised using Microsoft Excel 2013 software. The data was then subjected to calculations to determine Pearson's linear correlation coefficient. To ensure that the correlation determined was not random, the 2-tailed test was applied considering a significance interval (a) of 0.05 %.

The statistical calculations and graphs were drawn up using Origin Pro 8 software.

4.5 Determining the deposition rate of the PVD-PECVD system.

With the film thickness measurements from the sample characterisation stage, previously presented in subsection 4.3, the average deposition rate of Ag-NPs from the PVD-PECVD technique can be determined. To do this, the ratio between the film thickness and the time taken to produce it was calculated. The results were expressed in [im/min.

4.6   Microbiological test.

Once the film production and characterisation stages had been completed, the samples were sent for microbiological testing to study the antimicrobial effect produced in the tissues by the deposition of Ag-NPs. To do this, different samples coated with Ag-NPs were placed in 16 60 X 15 mm Petri dishes seeded with S. *aureu* (INCQS 00039, ATCC 6538) and *C. albicans* (Robin, ATCC 64124), both at a concentration of 9×10^8 cells/mL. The purpose of this mixed culture was to help assess the microbicity of the samples in real situations in which the tissues would come into contact with fungi and bacteria simultaneously. The cells were grown in 5 mL of tryptone soya agar (TSA) culture medium. The culture plates were incubated at 37° C for 7 days with the coated sides of the quintuplicates from each group facing the culture medium. The duration of this experiment was determined on the basis of a pilot test in which it was observed that the cultures, inoculated under the conditions of this experiment, showed no changes after 7 days in contact with the samples. At the end of this period, images of the Petri dishes were recorded with a photo-documentarian (Alpha Innotech) for later assessment of the results.

The antimicrobial effect was confirmed when the colony growth was limited to the edge of the samples; in the other cases where the cultures proliferated on the films, no antimicrobial effect was found. At the end of this test, the set of samples that showed the greatest effectiveness in inhibiting the microorganisms was submitted to the test to determine the antimicrobial effect, as described in the next subsection.

4.7   Test to determine the antimicrobial effect.

The set of samples that showed the greatest efficiency in microbial inhibition, as studied by the microbiological test in subsection 2.5, was submitted to the antimicrobial effect test. This test was designed to ascertain whether the Ag-NPs caused cell death or microbiostasis. The most efficient group in terms of microbicity was sterilised in a Braslact vertical model autoclave at 120 °C, 1 atm for 15 minutes. The group was then immersed in 5 mL of 10 % saline solution, contaminated with 9×10^8 cells/mL of *S. aureus* and *C. albicans*. At the end of this stage, the samples were placed in Petri dishes containing 5 mL of sterile TSA medium and incubated for 5 days at 37 °C.

Microbicity was found in cases where there was no growth of microbial cultures; in other cases, microbiostasis was found.

4.8   Comparison between MS and PVD-PECVD techniques.

After characterising the samples, determining the deposition rate of the PVD-PECVD system and the antimicrobial effect obtained, the PVD-PECVD and MS deposition techniques were then compared in terms of Ag-NPs concentration, film thickness, Ag-NPs size and antimicrobial effect, with the aim of answering the following questions:

- null hypothesis (Ho): there are no statistically significant differences between films produced by PVD-PECVD or Magnetron Sputtering.

- Alternative hypothesis (Ha): there are statistically significant differences between films produced by PVD-PECVD and Magnetron Sputtering.

The results produced by this study were compared to those obtained by three authors who conducted experiments under similar conditions to this work, but using MS systems: Mejía et al. (2010), Chadeau et al. (2010a), Wang et al. (2007a). The statistical calculations and graphs were drawn up using Origin Pro 8 software.

CHAPTER 5

Results and discussion.

The results and discussion of this study were separated into three blocks: production, sample characterisation, and comparison between the PVD-PECVD and MS techniques (Figure 1). With regard to the first block, which only covered subsection 3.1, evidence was presented to support the hypothesis that it is feasible to produce Ag-NPs films by PVD-PECVD. The analysis of the second block provided the theoretical basis for the objective of characterising the samples produced by PVD-PECVD. With regard to the third block, which comprised subsections 3.5 to 3.7, the antimicrobial effect of the films was presented and discussed, and the characteristics of the coatings were compared between the PVD-PECVD and MS techniques. The analysis of this block provided the theoretical basis for the third objective of this study: to compare the microbicity and efficiency of samples produced by PVD-PECVD and MS.

5.1 Morphological analysis of tissue fibres.

Once the film production stage had been completed, as described above in subsection 4.2, it was observed that the initially white samples had become silver in colour. Naturally this change in colour was due to the presence of Ag-NPs in the fabrics, but this fact was only confirmed after reading the EDX spectra of these groups, which showed prominent peaks corresponding to silver atoms (Figure 12 A), not present in the samples in their natural state (Figure 12 B). Also corroborating this finding is the comparison between the SEM images of the samples that were not subjected to the PVD-PECVD process (Figure 13 A) and those that were (Figure 13 B). This comparison highlights the presence of various silver clus- ters on the surface of the samples submitted to the process, thus leaving no doubt as to the ability to produce Ag-NPs coatings on textiles using the PVD-PECVD technique.

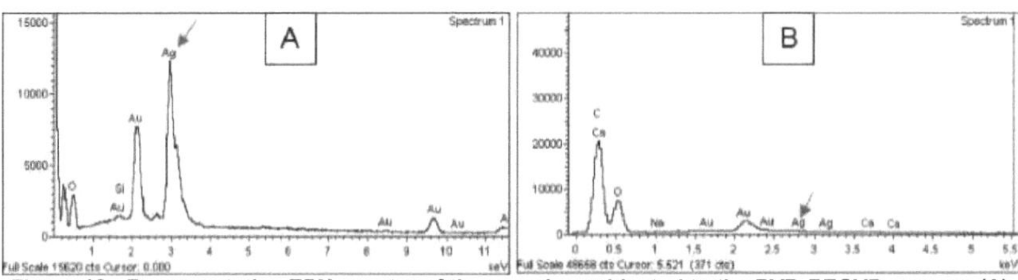

Figure 12 - Representative EDX spectra of the samples subjected to the PVD-PECVD process (A), and the samples in their natural state (B). The red arrows indicate the peak corresponding to silver.

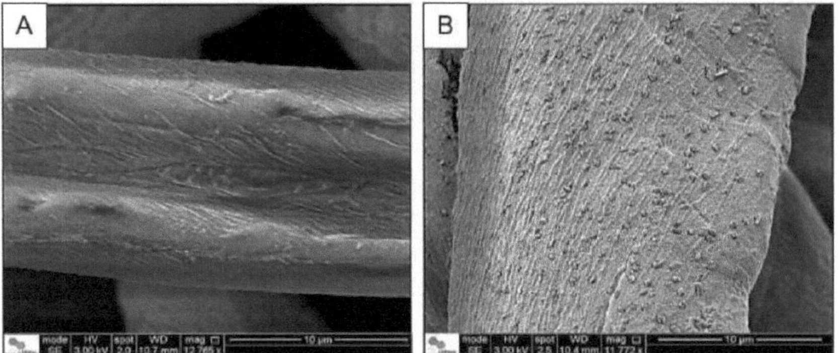

Figure 13 - Representative regions of a fabric fibre covered by the film: (A) border region between the solid film and the uncoated substrate; (B) solid film region.

5.2   Characterisation of coatings.

Once the ability to produce Ag-NPs coatings on textiles by PVD-PECVD had been realised, an understanding of the physical and chemical characteristics of this material was necessary in order to fully meet the proposal of this study. Subsection 5.2 has been subdivided into three parts where the morphological characteristics (subsection 5.2.1), physical properties (subsection 5.2.2) and chemical properties of the films produced (subsection 5.2.3) will be discussed.

5.2.1   Morphology.

Once the ability to produce Ag-NPs coatings on textiles by PVD-PECVD had been realised, it was necessary to understand the morphology of the films in order to fully meet the purpose of this study. For this reason, the physical analysis of the samples showed that the size of the clusters varied according to their distance from the solid film, as the more distant/isolated clusters were larger than those closer to the coating (Figure 14 A). It was also observed that the size of the clusters and the Ag-NPs that make them up depend on the type of sub-process by which the nanoparticles are ejected from the ASP. Finally, it was observed that the simple frequency with which Ag-NPs occur in films is related to the size of the nanoparticle. To make it easier to discuss these ideas, the first observation was labelled "observation -1", the second was labelled "observation - II", and the third "observation III". With regard to observation I, it can be said that the phenomenon described was due to the principle of conservation of mass/energy, since the clusters arose from the collision of atoms recently ejected from the ASP with others already deposited on the substrate. As this process is continuous throughout the deposition of the film, the energy resulting from the collisions accumulated in the clusters until it reached the melting point of the silver. As a result, the masses of the clusters were redistributed equally over the surface of the fabric, forming a film (Figure 14 B) in a manner very similar to what occurs in MS systems (MEJÍA et al., 2010; KELLY; ARNELL, 2000a). This phenomenon can be more easily understood by observing an ice cube melting on a flat surface (Figure 15 A), since the phase change observed in this material stems from the increase in the kinetic energy of the water molecules, liquefying the ice cube and causing the uniform redistribution of its mass (Figure 15 B), just as occurs with Ag-NPs clusters in the formation of films.

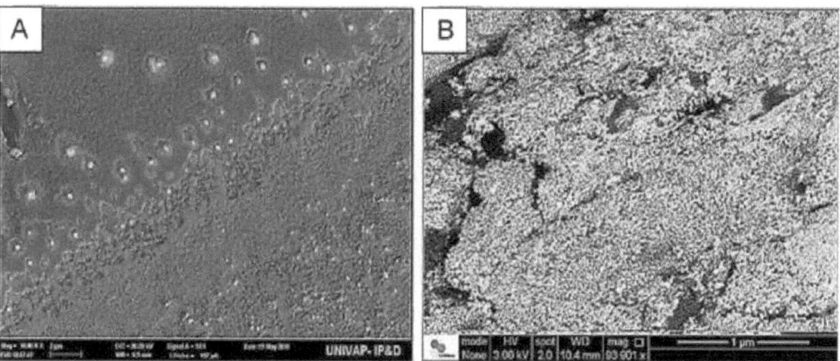

Figure 14 - Representative regions of a fabric fibre covered by the film: (A) border region between the solid film and the uncoated substrate; (B) solid film region.

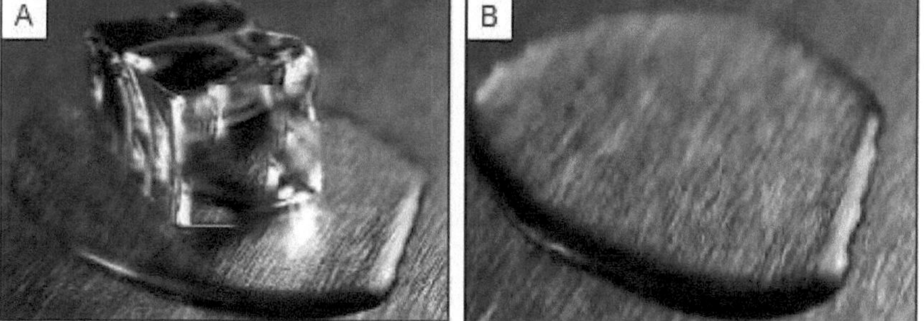

Figure 15 - Ice liquefaction process: (A) solid ice melting; (B) water in liquid phase.

Other evidence supports this explanation, such as the fact that the substrates of the 200 W/5-10 min groups showed points of carbonisation (Figure 16 A,B,D,E), even though they were electrically insulated and kept at a temperature ranging from 18-50^2 C during the deposition of the films - see subsections 4.1 and 4.2 -. Considering that the samples were not exposed to any exogenous heat source during film production, the only probable cause of this phenomenon would be the release of excess energy accumulated in the clusters produced under these parameters, something that did not occur at the other powers and deposition times.

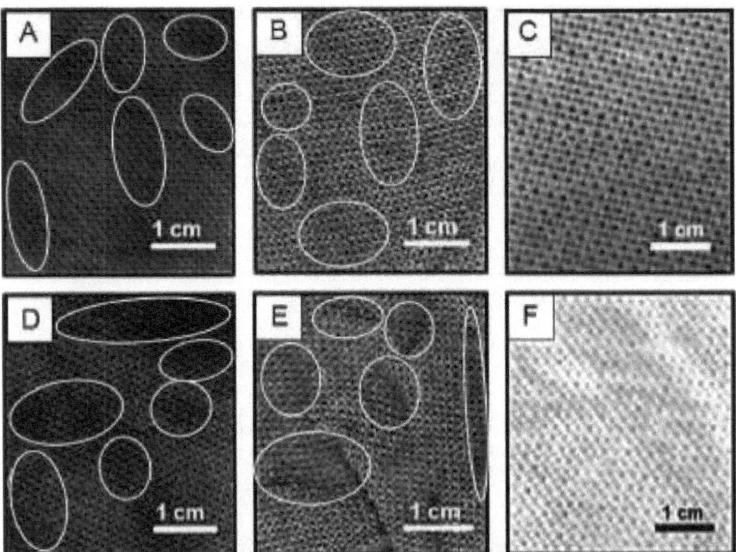

Figure 16 - Representative images of the groups with and without carbonisation points: 200W 10 min (front (A), back (D)), 200W 5 min (front (B), back (E)), other groups (front (C), back (F)). The presence of white circles highlights charred regions in samples (A)/(D),(B)/(E), not present in sample (C)/(F).

With regard to observation II, it can be said that the variation in Ag-NP diameters is also related to the deposition sub-processes resulting from the PVD-PECVD technique: sputtering and evaporation. Nanoparticles produced by sputtering have higher energy and smaller grain size compared to those produced by evaporation, which have Ag-NPs with lower energy and larger size (OHRING, 1992; OHRING, 2002). As the plasma sheath region around the ASP extends approximately 60 mm from its surface, the portion of the samples facing the centre of the ASP received more particles from the evaporation process, while the parts facing its edge received more particles from the sputtering process. Based on this principle and the "ice cube" effect, the uniformity observed in the coatings is understood to be the result of both the aggregation of the isolated Ag-NPs by the melting clusters and the redistribution of their mass in equal parts across the substrate.

With regard to observation III, so far various pieces of evidence have shown that the size of the clusters, and consequently that of the Ag-NPs, is not the same throughout the coating. However, not only did the diameter of the nanoparticles vary as a function of distance from the solid film, but also the simple frequencies with which each type of Ag-NP occurred in the coatings (Figure 17).

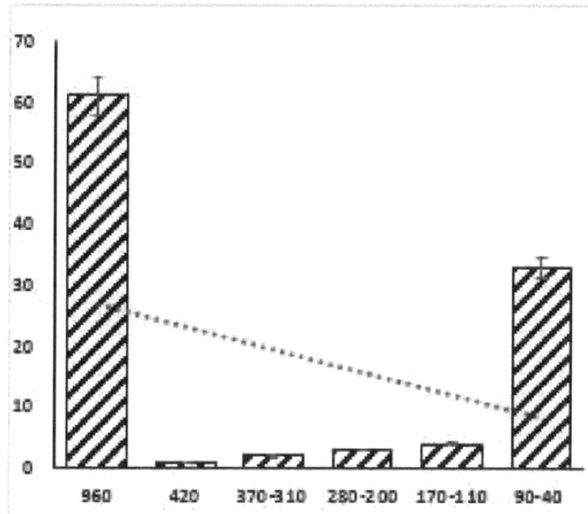

Figure 17 - Distribution of frequencies and diameters of Ag-NPs produced by PVD-PECVD.

In general, the Ag-NPs present in the films can be divided into three groups according to their diameters: group (1), relating to 960 nm Ag-NPs, corresponds to 61% (± 8.8%) of the particles present in the film; group (2), relating to Ag-NPs limited to the 110-420 nm range, corresponds to 6% (± 0.4%) of the particles present in the film; group (3), relating to Ag-NPs limited to the 40-90 nm range, corresponds to the remaining 33% (± 7.9%) of the particles that make up the entire (100%) coating. In this sense, the large distance between the diameters that make up each group corroborates the position of Kelly and Arnell (2000a) and Mejía et al. (2010), adapted to the PVD-PECVD technique, that the Ag-NPs were ejected from the ASP by means of the sputtering and evaporation sub-processes. According to the definition of these sub-processes, the Ag-NPs produced by evaporation certainly corresponded to group (1). However, the definition of the Ag-NPs produced by sputtering was not very precise, as it was not possible to clearly determine whether the nanoparticles, limited to diameters of 40-90 nm, were effectively ejected from the ASP in this dimension, or whether they were smaller and grew due to the fusion of the clusters. This doubt is further supported by the divergence between the minimum diameter of the Ag-NPs produced by ASP- PECVD and the diameters usually found in the literature for films produced by MS (KELLY; ARNELL, 2000a; MEJÍA et aL, 2010). With this in mind, the present study proposes that the smaller clusters formed by particles ejected by sputtering received matter from the larger clusters formed by particles from evaporation, and found the equilibrium point between their redistributed dimensions at 40 nm. This effect occurred in a similar way to two ice cubes of different sizes kept in contact, which level out in terms of volume when the water resulting from their liquefaction mixes. If this volume of water loses enough energy to solidify, a single ice stone is formed whose dimensions equal the average of the dimensions of the cubes that preceded it. As a result, as the Ag-NPs in group (1) melt, the Ag-NPs in group (3) become more frequent in the coatings. The trend curve

31

shown in figure 17 also corroborates this theory. It shows that from the Ag-NPs in group (1), there is a progressive and almost regular shift towards the formation of Ag-NPs in group (3). Therefore, the nanoparticles included in group (2) actually represent transition phases from the Ag-NPs of group (1) towards group (3).

5.2.2   Physical properties.

Sample properties such as thickness, wettability and roughness, and the effect of inhibiting the growth of micro-organisms, were a direct result of the PPAg deposited on the substrates. This variable numerically expressed the amount of silver deposited on each film by means of the intensity of the peaks corresponding to silver present in the EDX spectra of the samples. Observation of these spectra showed that the silver peaks grew as the deposition time and/or plasma power increased (Figure 18). As a result, it was observed that the 200 W/10 min group (Figure 18 A) showed on average higher silver peaks than those found in the 150 W/10 min groups (Figure 18 B), which in turn showed higher peaks than those obtained in the 100 W/10 min (Figure 18 C) and 75 W/10 min (Figure 18 D) groups, with the latter group showing the lowest silver peaks of the entire experiment.

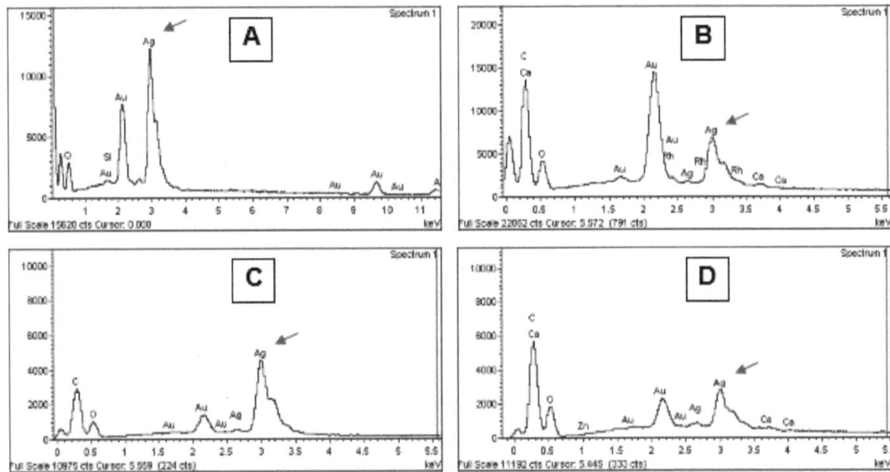

Figure 18 - Representative EDX spectra of the samples produced by 10-minute depositions: (A)200W, (B)150W, (C)100W, (D)75W. The red arrows indicate the position corresponding to the silver peak in each spectrum.

Analysing the values expressed by the EDX spectra (Table 3), it was observed that the PPAg present in the samples did not vary significantly, since it went from a PPAg of approximately 0% on average for the 75 W/0.5 min group, to a PPAg of approximately 17% on average for the 200 W/10 min group. Regardless of this, this result suggests that there is a correlation between the PPAg in the films, the duration of the depositions and the plasma power.

Table 3 - PPAg of the samples, considering the intersections of the "time" and "power" variables.

Power (W)	Time(min)			
75	0(± 0)	0 (± 0,22)	0,07(± 0,004)	0,3 (± 0,1)

100	0(± 0)	0,6 (± 0,02)	0,7 (± 0,05)	1,1 (±0,17)
150	0 (± 0,2)	2,1 (± 0.005)	3 (±0,08)	8 (± 0,5)
200	0(±0,1)	2,3 (± 0,02)	5 (± 0,3)	17 (±1)

In fact, determining Pearson's correlation coefficient indicated that the correlation observed between these variables was extremely strong, since all the correlation coefficients (r) calculated showed values above 0.9 (Table 4). The only exception was the groups produced for 0.5 min and the 75W/1 min group, for which the value of (r) could not be determined. The calculated "r" was also subjected to a two-tailed hypothesis test, which ensured its significance at a level of 95%, indicating that the correlation observed was not random. As the concept of power in its simplest expression is the amount of work done by a force as a function of time, the association observed between plasma power and PPAg was expected even in relation to the negligible amount of silver deposited in the processes carried out during 0.5 min and 75W/1 min. However, this information cannot be statistically verified as these groups showed PPAg values very close to 0%. The low content of Ag-NPs in these samples was due to the insufficient quantity of silver atoms ejected from the ASP to form a film at appreciable levels. Thus, the presence of the scarce Ag-NPs on the surfaces of the 0.5 min groups and the 75W/1 min group occurred in an unpredictable and very isolated manner in different regions of the substrates. With regard to the other groups, the Ag-NPs contribution was large enough to form a film, which resulted from the progressive depletion of free spaces on the sample surface as the deposition time and/or plasma power increased.

Table 4 - Pearson's correlation coefficient between the variables "time" and "power".

Power (W)	Time(min)			
75	Nil	Nil	0,96342	0,9626
100	Nil	0,96484	0,99998	1,00
150	Nil	1.00	0.96579	0,96484
200	Nil	0,96579	1,00	0,99998

The thickness of the coatings (Table 5) varied from 0.3 pn (± 0.02 ^im) to 12 [jm (± 0.94 [im) depending on the parameters used. The growth profile of this variable (Figure 19 A) followed that of PPAg (Figure 19 B), since the accumulation of Ag-NPs in overlapping strata is what makes up the film. Consequently, the determining parameters in the PPAg, added to the reactor's Ag-NPs deposition rate (Table 6), are equally important for understanding the growth in coating thickness (Figure 19 C).

Table 5 - Layout of the thicknesses of the films deposited on the samples, considering the crossings of the "time" and "power" variables.

Power (W)	Time(min)			
75	0(± 0)	0(± 0)	0,3 (± 0,02)	0,5 (± 0,03)
100	0(± 0)	0,6 (± 0,03)	1,4 (± 0,05)	1,7 (± 0,07)
150	0(± 0)	0,8 (± 0,04)	2,3 (± 0,03)	3 (±0,17)
200	0(± 0)	7,7 (± 0,4)	9,7 (± 0,6)	12 (± 0,94)

Table 6 - Breakdown of the system's deposition rates, considering the crosses of the "time" and "power" variables.

Power (W)	Time(min)			
75	0(± 0)	0(± 0)	0,06 (± 0,034)	0,.05 (± 0,007)
100	0(± 0)	0,6 (± 0,05)	0.28 (± 0,058)	0,17 (± 0,01)
150	0(± 0)	0,8 (± 0,05)	0,46 (± 0,09)	0,3 (± 0,091)

| 200 | 0(± 0) | 7,7 (± 1,02) | 1,94 (± 0,087) | 1,2 (± 0,085) |

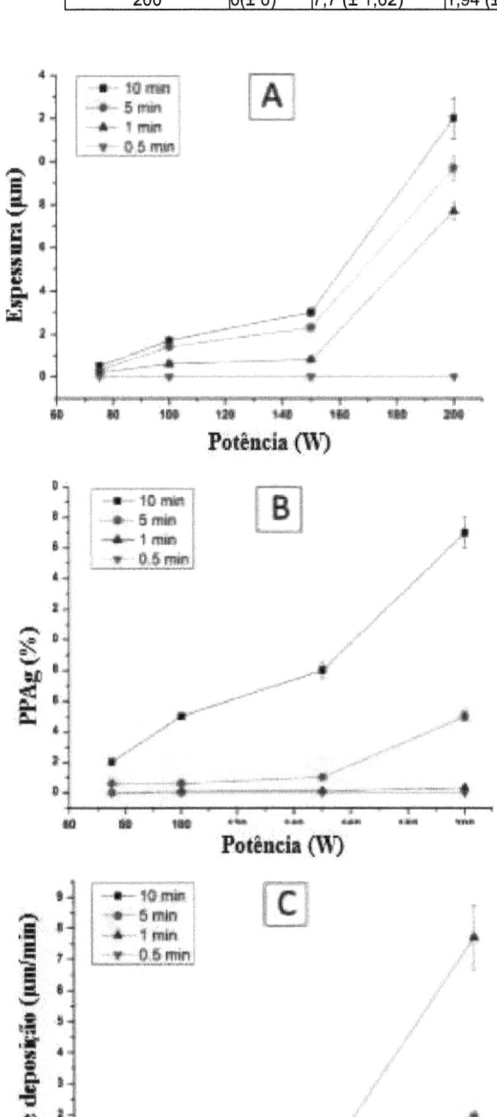

Figure 19 - Characteristics of the films and the process: (A) thickness of the coatings, (B) PPAg in the samples, (C) deposition rate of the PVD-PECVD system.

So far the causes of variation in the PPAg deposited on the samples, the thickness of the films and the deposition rate of the PVD-PECVD system have been presented. Observation of these causes reveals a conflict in determining the maximum deposition time that the system could support, since the deposition rate, which varied from 0.2 μm/min ($\pm$ 0.04 [im/min) to 1.2 μm/min ($\pm$ 0.085 μm/min), grew inversely to the duration of the depositions, while the PPAg grew proportionally to this variable. Two observations emerged from this analysis: the first, relating to the growth of the deposition rate, suggests the existence of a maximum time limit after which deposition would cease; the second, relating to the growth of the PPAg, suggests that the "deposition time" parameter is the most important for achieving coatings with previously determined thicknesses. Although the maximum deposition time was not the subject of this study, empirical experience working with PVD-PECVD systems suggests that the electrical discharge would not be stable for much longer than 10 minutes due to the appearance of metastable arcs. This phenomenon results from the increasing conductivity of the sample fixing system as the process progresses, since a thin layer of Ag-NPs is deposited on the ceramic disc (Figure 20 A) and the glass slide (Figure 20 B) - compare Figure 20 B and Figure 20 C - concomitantly with the formation of the films on the fabrics. As a result, electrically conductive islands appeared in these dielectric materials, which began to act as a second anode, disturbing the stability of the plasma in the system.

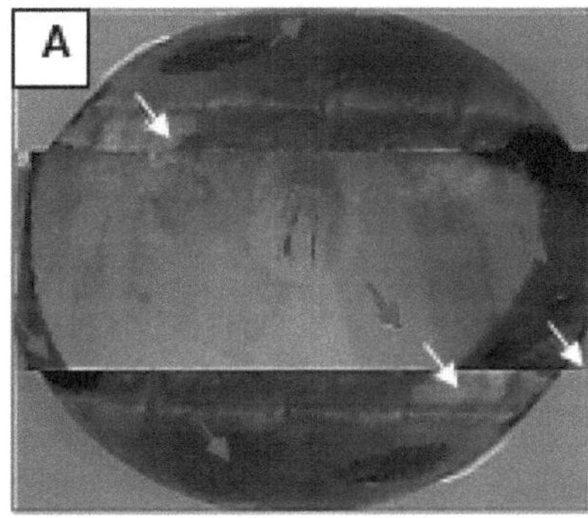
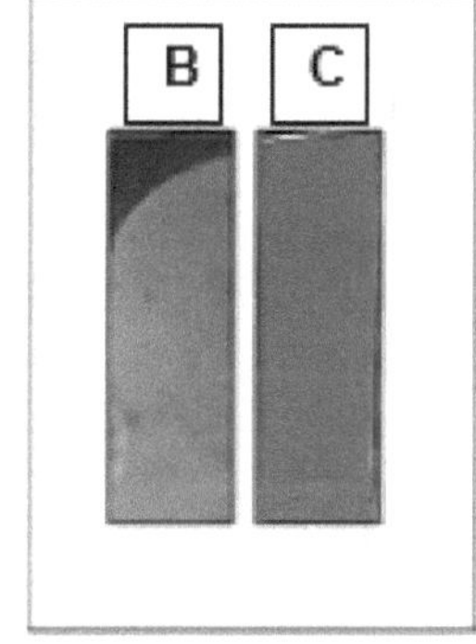

Figure 20 - Components of the sample isolation and fixation system: (A) ceramic disc with glass slide resting on its central portion; (B) glass slide covered with silver film; (C) glass slide in its original state. The red arrows in figure 20 A indicate portions of the disc covered with the film, while the white arrows indicate uncovered regions.

With regard to the wettability of the sample surface, the analysis of the static CA of the water droplet (Table 7) revealed that the groups produced in 0.5 min depositions, for all powers, and the group produced during 75 W/1 min, showed statistically equivalent CAs to the control group not covered by the film. On the other hand, it was observed that the CA of the other groups decreased as the PPAg increased in the samples. Consequently, the 200 W/10 min group had a lower CA (89.3°±4.5°) than the others. The inverse growth profile between the PPAg and CA variables has been observed in Ag-NPs films deposited on cotton fabrics produced by different techniques (ZILLE et al, 2015; VU et al, 2013). In general, the scientific community considers the primary cause of this effect to be the increased roughness of the substrate due to the different sizes of Ag-NPs in the film (ZILLE et aL, 2015; LI et aL, 2008; LEE, 2003; DURAN et aL, 2007). In this respect, the present study confirmed this premise, since the deposition of Ag-NPs increased the roughness of uncoated substrates (Figure 22 A) by up to 110.5% (Figure 22 B).

Table 7 - Distribution of the water contact angle values measured on the samples, taking into account the crossing of the "time" and "power" variables.

Power (W)	Time (min)			
75	142,6 (± 3,1)	142,6 (± 5,7)	125,9 (± 5,3)	116,9 (± 4,5)
100	142,6 (± 5,1)	131,9 (± 4,6)	125,7 (± 4,9)	113,2 (± 5,6)
150	142,6 (± 5,8)	125,7 (± 5,3)	118,9 (± 4,9)	110,9 (± 4,8)
200	142,6 (± 6,0)	110,9 (± 6,5)	110,6 (± 4,5)	89,3 (± 4,5)

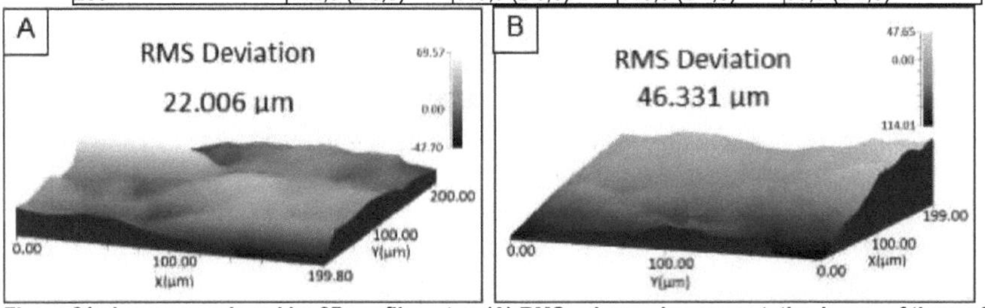

Figure 21 - Images produced by 3D profilometry: (A) RMS value and representative image of the surface of the groups not coated with Ag-NPs; (B) RMS value and representative image of the groups coated with Ag-NPs.

5.3 Comparison between PVD-PECVD and MS techniques.

With the films readily produced and characterised in terms of their properties and limitations, the discussion of the results relating to the antimicrobial effect produced by the Ag-NPs deposited on the substrates can be developed, as will be shown in sub-section 5.3.1. Having completed the presentation and discussion of this stage, sub-section 5 will conclude with the discussion relating to the statistical comparison between the PVD-PECVD and MS techniques, carried out in sub-section 5.3.2.

5.3.1 Antimicrobial effect produced by the film.

The results of the microbiological test showed that there was no growth of the culture on the samples produced at 150 W and 200 W for the times of 1 min, 5 min and 10 min, which were therefore considered effective in terms of inhibiting the growth of microorganisms. In all other cases, the samples were partially or completely covered by the culture and were therefore ineffective in terms of their microbicidal properties (Figure 23). With regard to the effect observed, these results are in line with published studies involving not only S. aureus and C. albicans, but also Escherichia coli and Klebsiella pneumoniae (EGBUTA; MWANZA; BABALOLA, 2017b; MITRA et aL, 2017b; LEE, 2003; DURAN et aL, 2007; PERELSHTEIN et aL, 2008).

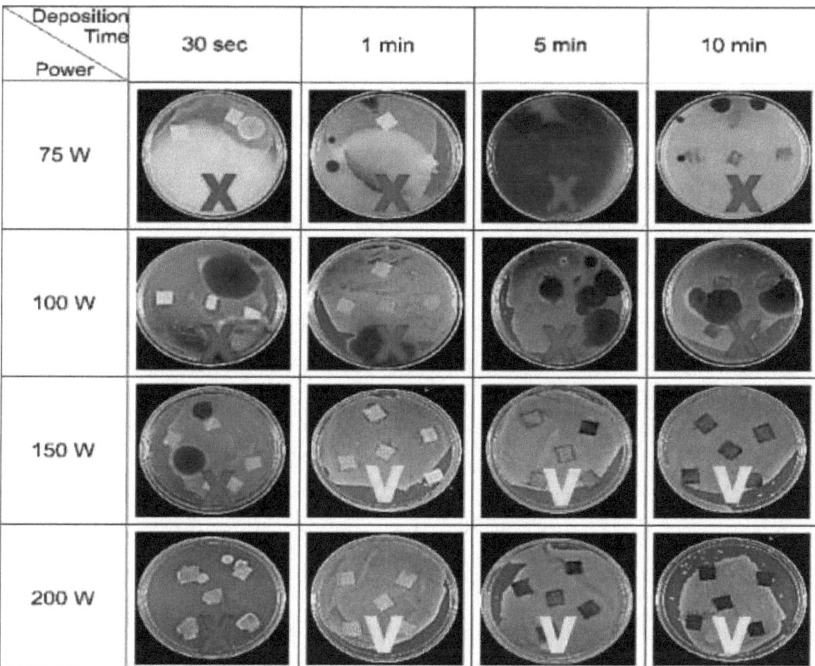

Figure 22 - Antimicrobial test. The culture medium is partially covered with S. aureus, as seen by the milky appearance, and C. albicans, as seen by the dark or white regions with a tangled appearance. The samples correspond to square fragments of tissue and are marked with a red "X" when they were ineffective in inhibiting the micro-organisms, and with a yellow "V" when they were effective.

The antimicrobial effect achieved by Ag-NPs is closely associated with the physical characteristics of the coatings. In this sense, the concentration of 2.1% (± 0.005%) of Ag-NPs in the samples marked the minimum limit at which efficacy was achieved. Analysing this result shows that, as reported by Budama et al. (2013), Ag-NPs demonstrate antimicrobial efficacy even in small concentrations.

quantities.

With regard to the diameter of the Ag-NPs, several studies have highlighted the microbicidal power of nanoparticles ranging from 1 nm to 100 nm in diameter (RADETIc, 2013; LEE, 2003; DURAN et al., 2007; BUDAMA et al., 2013; DUBAS; KUMLANGDUDSANA; POTIYARAJ, 2006). In this respect, the present methodology corroborates the literature since 33% (± 7.9%) of the Ag-NPs present in the coatings are limited to the 40-90 nm diameter range. However, there have been no studies demonstrating the effects of particles larger than this range and neither can we say, through this experiment, which particles in the coatings were or were not responsible for bacterial and fungal inhibition, since 67% (± 7.9%) of the Ag-NPs counted by the IMAGE J software had a diameter of more than 100 nm and less than 960 nm.

Despite these associations, various studies have linked the anti-microbial effect of Ag-NPs to the thickness of the silver films. However, there is no consensus regarding this property in terms of efficacy, as some studies, such as that by Sataev et al. (2014), show microbicidal coatings of 0.5-0.6 [im, while other studies, such as those by Chadeau et al. (2010b) and Wang et al. (2007a), have achieved the same efficacy with extremely thin silver coatings of 0.001 um to 0.4 [im. The combination of these findings supports the premise that Ag-NPs coatings are microbicidal within a wide range of thicknesses, which in the case of this study had a lower limit of 0.8 um (± 0.04 [im), and an upper limit of 12 [um (± 0.94 [im). In contrast to the findings in the literature (CHADEAU et aL, 2010b; WANG et aL, 2007b; MEJÍA et aL, 2010), the present methodology also produced coatings 0.3 um (± 0.02 [im) thick with no microbicidal effect detected. These results therefore suggest that the association between microbicity and coating thickness is merely fortuitous.

Finally, according to Chadeau et al. (2010b) Ag-NPs films deposited on cotton fabrics by MS achieved a significant reduction in the number of colony forming units (CFU) after 24 hours. On the other hand, still with regard to the MS technique, Mejía et al. (2010) produced totally antimicrobial samples from depositions of 1 min in duration. Compared to these findings, the PVD-PECVD method proved to be more efficient, as at no point in the experiment was there any growth of micro-organisms on films produced after 1 min, at 150 W and 200 W power.

Due to the high PPAg of the groups produced at 200 W power for 10 minutes, and the complete absence of micro-organisms present in these samples, this sample set was used in the test to determine the antimicrobial effect. The results showed no C. albicans colony growth in both the group without Ag-NPs (Figure 24 A) and the group with Ag-NPs (Figure 24 B). However, both samples showed growth of the S.aureus culture in the TSA medium, with the colony growing on the samples without Ag-NPs and around the samples with Ag-NPs, which showed a narrow inhibitory halo. This result shows that the antimicrobial effect caused by this methodology is bacteriostatic and fungicidal.

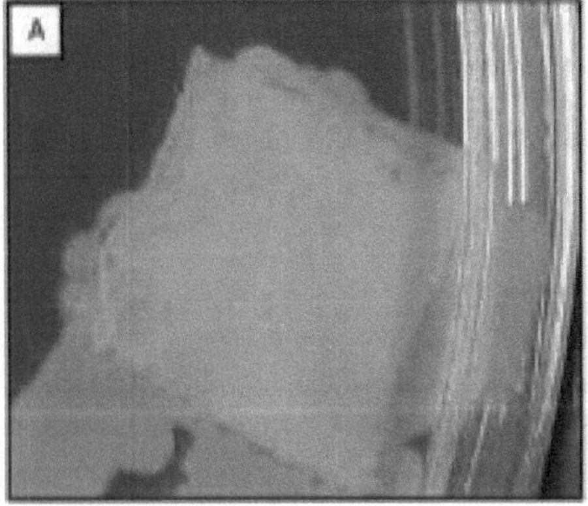

Figure 23 - Test to determine the antimicrobial effect: (A) representative image of the quintuplicate of the samples without Ag-NPs; (B) representative image of the groups coated with Ag-NPs. The red arrow in figure (B) shows the formation of a narrow inhibitory halo that was not observable in (A).

To date, few studies have related the microbicidal effect observed on textiles coated with Ag-NPs to the wettability and roughness of the fabrics (ZILLE et al., 2015; VU et al., 2013). However, there are studies showing that surfaces with high RMS have numerous peaks and valleys, which provide favourable sites for eukaryotic cell adhesion (SANTOS et aL, 2016; A MARTIN PJ, 2007). As the chemical composition of the cell membrane is essentially the same

in all living beings, it is reasonable to assume that the same effect would occur in relation to bacteria and fungi. However, this study contradicts this logic because the groups with the highest RMS values were the most antimicrobial. As a result, the direct relationship between the roughness quantities and the antimicrobial effect is coincidental, since in reality the phenomenon observed stems from a change in the CA of the samples, a variable dependent on the RMS value. Therefore, the incorporation of Ag-NPs into the substrates makes them more hydrophilic and consequently more repulsive to hydrophobic structures such as the plasma membrane/cell wall of microorganisms. This hypothesis effectively ties in with and corroborates the results of the antimicrobial effect test, which revealed that the films produced by this methodology are ineffective at killing colonies, except in cases of prolonged bacteriostasis (SUN; WANG; YAN, 2017), but are highly effective at preventing infections by preventing the attachment of microorganisms (LEE; YEO; JEONG, 2003; DURAN et al, 2007). However, there is no consensus on what effectively inhibits microbial growth, with several studies associating this effect with the inherent microbicity of silver, or its gradual and persistent release from the tissue (CHADEAU et al., 2010b; PERELSHTEIN et al., 2008; SUN; WANG; YAN, 2017).

5.3.2 Statistical comparison between the PVD-PECVD and MS techniques.

In order to fulfil the third objective of this study, the PVD-PECVD and MS systems were compared in terms of the characteristics and effects of the films produced. To provide the necessary data for this comparison, as described in subsection 4.8, three studies of silver coatings deposited on textiles using MS were selected. Once all the necessary information was available, a table was drawn up showing the measured values of each determining variable in these experiments, depending on the technique used (Table 8). Variables that had not been studied by at least one author of the MS technique were discarded from this table, as in these cases no information regarding the comparison between the PVD-PECVD and MS techniques would be obtained.

Table 8 - Comparative distribution of the values of the variables of the films produced by PVD-PECVD and MS. Empty cells correspond to data not made available by the author in his work.

Parameters	PVD-PECVD system	MS System: Experiment by M.I.Mejía et al.(2010)	MS System: Experiment by E Chadeau et al. (2010)	MS System: Experiment by Wang, Hongboand et al. (2007)
Minimum time to produce microbicidal samples (min)	1	1	-/-	5,7
Power (W)	150	60	-/-	40
Pressure (Torr)	0,03	0,0075	0,0015	0,0160
PPAg(%)	2,1	23,36	-/-	-/-
Intensity of the microbial effect (%)	100	100	100	100
Minimum thickness of microbicide films ([im)	0,8	0,04	0,001	0,002
Maximum size of Ag-NPs (nm)	960	9,8	-/-	-/-
Minimum size of Ag-NPs (nm)	40	4	-/-	-/-

Table 6 shows that: the minimum thickness of the microbicide films is on average 98% greater in films produced by PVD-PECVD than by MS; the minimum PPAg of the microbicide films is on average 91% greater in films produced by MS than by PVD-PECVD; the diameter of the nanoparticles of the microbicide films is on average 99% greater in films produced by PVD-PECVD than by MS. As the differences found between the characteristics of the films were quite significant, there was no need to employ any hypothesis testing to ensure that the production of Ag-NPs films on textiles by PVD-PECVD would result in coatings with characteristics quite different from those produced by MS.

With regard to the means by which Ag-NPs can be deposited on textiles, both the PVD-PECVD technique and MS have demonstrated the ability to produce equally efficient coatings in terms of microbicity. Therefore, in this respect, it was not possible to determine any advantages of one system over the other, based solely on the present results and the studies available in the literature on the microbicity of Ag-NPs deposited on textiles by MS (CHADEAU et al., 2010b; WANG et al., 2007a; MEJÍA et al., 2010). However, although MS systems require a working pressure reduced by 75% (MEJÍA et al., 2010), 95% (CHADEAU et al., 2010b) or 47% (WANG et al., 2007b) compared to the PVD-PECVD system, in general the MS technique uses less energy and resources to produce microbicide films. According to Mejía et al. (2010), at a power of 60 W the MS system deposits a mass of Ag-NPs 91% greater than the PVD-PECVD system at a power of 150 W. As a result, considering an arbitrary electricity tariff value, a saving of 0.01 US$ - or 150 % - was estimated per process operated on MS systems instead of PVD-PECVD. However, this observation does not disqualify the PVD-PECVD technique in comparison to MS, but rather highlights the need to better exploit the properties of this system since its versatility, which is far superior to that of the MS technique, makes it easy to overcome its disadvantages by making it possible to: diversify plasma configurations, use different precursor gases alone or concomitantly, modify the method of fixing the samples, use ASP composed of silver alloys or other metals, readjust the dimensions of the ASP, among other modifications that would optimise the efficiency of PVD-PECVD systems. It should also be emphasised that the total duration of the PVD-PECVD process is around 2 minutes, or 10 % less than the MS process. This difference is a consequence of the ideal working pressure of the MS system being lower than that of the PVD-PECVD system, so MS systems require a longer time to reach the ideal conditions for depositing the film, and for reopening the chamber at the end of the process. This undoubtedly gives the PVD-PECVD system an important technical advantage, as working at higher pressures saves investment in the vacuum and sealing system, and reduces the total duration of the process. As time is a decisive factor in large-scale production, even though the PVD-PECVD technique does not surpass the efficiency of MS systems, this technique is an interesting option for industrial use in processes that require rapid production and continuous product innovation. This versatility is especially interesting for industry, which inevitably needs to continually innovate and improve its products in order to remain competitive in the market.

CHAPTER 6

Conclusion.

The PVD-PECVD technique has only been used to promote doping and deposition of more uniform carbon coatings than those obtained industrially by carburisation, nitriding and carbonitriding techniques. However, although this technique has promising characteristics for the deposition of Ag-NPs on fabrics, no study has yet been carried out to verify them. For this reason, this study had three objectives:

I) to verify the feasibility of producing Ag-NPs coatings on cotton fabrics using the PVD-PECVD technique;

II) determine the physical and chemical properties and the antimicrobial effect of the films produced by PVD-PECVD;

III) to compare the results produced by the PVD-PECVD technique with the results available in the literature for the MS technique.

With regard to objectives (I) and (II), the results lead to the conclusion that the PVD-PECVD technique is an effective alternative for producing cotton fabrics with incorporated silver nanoparticles. The films produced show good uniformity regardless of the initial size of the nanoparticles ejected from the ASP. This characteristic results both from the aggregation of isolated Ag-NPs by the melting clusters and from the redistribution of their mass in equal parts on the substrate due to the "ice cube" effect. The size of the Ag-NPs deposited by this technique varies in diameter according to their distance from the solid film, and also according to the type of sub-process by which the nanoparticles are ejected from the ASP: sputtering or evaporation. The largest Ag-NPs produced by PVD-PECVD have a diameter of 960 nm, and decrease in size until they reach around 40 nm before fusing to the solid film. The simple frequency with which these particles occur in the coatings shifts in favour of the smallest Ag-NPs limited to the 40 nm - 960 nm range. With regard to the concentration of silver in the films, PPAg shows a maximum variation of 17% of the total mass of the sample. Among the variables configured in the system, the duration of the depositions and the plasma power stand out in determining the PPAg in the samples, and consequently in determining the other properties of the films. Finally, with regard to the antimicrobial effect, it was found that the deposition of small amounts of Ag-NPs, around *2.1% (± 0.005%),* makes the fabrics 100% effective in inhibiting micro-organisms. With regard to objective (III), the MS and PVD-PECVD techniques differ significantly from each other. The PVD-PECVD system is less efficient than MS and more limited in that it does not support prolonged depositions when dielectric materials are used to fix the position of the samples; on the other hand, the MS technique operates processes that take longer and are less versatile than PVD-PECVD. However, both techniques demonstrate equivalent efficiency in terms of the microbicity of the films produced.

The findings of this study certainly ensure that the PVD-PECVD technique is a valid alternative to the processes currently used to produce fabrics with incorporated Ag-NPs. However, until the disadvantages presented by this system are overcome, the MS technique will continue to be the best option. Therefore, this session concludes with the proposal of some studies that will certainly make a significant contribution to improving the PVD- PECVD technique, especially in the textile field:

- To study the characteristics of Ag-NPs films obtained by PVD-PECVD processes using other gases, such as oxygen and hydrogen, instead of argon;

- To study the characteristics of Ag-NPs films deposited by PVD-PECVD on other textile materials;

- To develop more efficient means of fixing and isolating the samples, and to study the effect of this modification on the deposition rate of the PVD-PECVD system;

- To study the microbicidal properties of films produced by PVD-PECVD using ASP composed of silver alloys or other metals known for their microbicidal effect, such as copper and gold;

- To study the effects of readjusting the dimensions of the ASP used in this study in relation to the properties of the films and the deposition rate of the system.

References

A MARTIN PJ, C. C. P. E. H. A. M. F. S. R. B. The mechanical and biocompatibility properties of DLC-Si films prepared by pulsed DC plasma activated Chemical vapour deposition. *Diamond and Related Materials,* v. 16, n. 8, p. 1616 - 1622, 2007. Accessed on 21/05/2016. Cited on page 52.

AL-MARIRI, A.; YOUNES, A. A.; SHARABI, N. Characterisation of thermophilic strepto- cocci isolated from rustic white cheese. *The Journal of General and Applied Microbio- logy,* v. 59, n. 2, p. 97 - 103, 2013. Cited on page 20.

ALTENBAHER, B.; TURK, S. Sostar; FIJAN, S. Ecological parameters and disinfection effect of low-temperature laundering in hospitals in Slovenia. 2011. Cited on page 20.

ALVES, L. R. A. et al. *The restructuring of cotton farming in Brazil.* 2006. Thesis (Doctorate) - University of São Paulo. Available at: <http://www.teses.usp.br/teses/ disponiveis/11/11132/tde-09112006-144525/>. Cited twice on pages 17 and 18.

ASTAPENKO, V. A. Contributions of Plasma Physics. *Contributions to Plasma Physics,* v. 53, n. 7, p. 507 - 509, agust 2013. ISSN 0863-1042. Cited on page 24.

AVCI, U. et al. Cotton Fiber Cell Walls of *Gossypium hirsutum* and *Gossypium barbadense* Have Differences Related to Loosely-Bound Xyloglucan. *PLoS ONE,* Public Library of Science, v. 8, n. 2, p. e56315 -, 2013. ISSN 1932-6203. Available at: <https://www.ncbi.nlm.nih.gov/pmc/articles/PMC3572956/>. Cited on page 17.

BAKI, M. E. et al. Effects of silver nano-particles on sperm parameters, number of Leydig cells and sex hormones in rats. *Iranian Journal of Reproductive Medicine,* Research and Clinicai Centre for Infertility, v. 12, n. 2, p. 139 - 144, 2 2014. ISSN 1680-6433. Available at: <https://www.ncbi.nlm.nih.gov/pmc/articles/PMC4009568/>. Cited on page 21.

BRINKER, C. J.; CLARK, D. E.; BALLANCE, J. B. Better Ceramics Through Chemistry III. *Materials Research Society,* Pittsburgh, p. 717 - 729,1988. Available at: <http://dtic.mil/dtic/tr/fulltext/u2/a207128.pdf>. Accessed on: 11/04/2018. Cited 3 times on pages 6, 22 and 23.

BUDAMA, L. et al. A new strategy for producing antibacterial textile surfaces using silver nanoparticles. v. 228, p. 489 - 495, 7 2013. Available from: <http://linkinghub.elsevier.com/retrieve/pii/S1385894713006347>. Cited 3 times on pages 21, 50 and 51.

CAPDAREST-AREST, N.; GONZALEZ, J. R; TURKER, T. Hypotheses for Ongoing Evolution of Muscles of the Upper Extremity. *Medicai hypotheses,* v. 82, n. 4, p. 452-456, 4 2014. ISSN 0306-9877. Available at: <https://www.ncbi.nlm.nih.gov/pmc/ articles/PMC4059683/>. Cited on page 16.

CASIRAGHI, C.; FERRARI, A. C.; ROBERTSON, J. Raman spectroscopy of hydro- genated amorphous carbons. *Phys. Rev. B,* American Physical Society, v. 72, Aug 2005. Available at: <http://link.aps.Org/doi/10.1103/PhysRevB.72.085401>. Cited on page 26.

CHADEAU, E. et al. Anti-Listeria innocua activity of silver functionalised textile prepared with plasma technology. *FOOD CONTROL,* v. 21, n. 4, p. 505 - 512, APR 2010. ISSN 0956-7135. Cited twice on pages 11 and 37.

CHADEAU, E. et al. Listeria innocua activity of silver functionalised textile prepared with plasma technology. v. 21, n. 4, p. 505 - 512, 2010. Available at: <http://www.sciencedirect.com/science/journal/09567135>. Cited 5 times on pages 16, 19, 51, 53 and 55.

CHAN, A. J.; OLIVEIRA, A. C. M. da C. *The effects of globalisation on the textile industry.* 2010. Master's dissertation. Available at: <http://hdl.handle.net/10438/5605>. Cited on page 16.

CHENG, S. et al. A PCR assay for Identification of Enterococcus faecium. 1997. Cited twice on pages 19 and 20.

CHIDAMBARAM, M.; KRISHNASAMY, K. Modifications to the Conventional Nanopreci- pitation Technique: An Approach to Fabricate Narrow Sized Polymeric Nanoparticles. *Advanced Pharmaceutical Bulletin,* Tabriz University of Medical Sciences, v. 4, n. 2, p. 205 - 208, 6 2014. ISSN 2228-5881. Available at: <https://www.ncbi.nlm.nih.gov/ pmc/articles/PMC3915822/>. Cited on page 21.

C.J.BRINKER et al. Fundamentals of sol-gel dip coating. *Thin Solid Films,* v. 201, n. 1, p. 97 - 108, 1991. ISSN 0040-6090. Available at: <http://www.sciencedirect.com/science/article/pii/004060909190158T>. Cited on page 23.

CLIFTON, I. J.; PECKHAM, D. G. *Defining routes ofairborne transmission of Pseudo- monas aeruginosa in people with cystic fibrosis.* 2010. Cited on page 19.

CRUZ, E. D. de A. et al. *Staphylococcus aureus and methicillin-resistant Staphylococcus aureus in workers at a university hospital.* 2008. Thesis (Doctorate) - University of São Paulo. Available at: <http://www.teses.usp.br/teses/disponiveis/ 83/83131/tde-06102008-151422/>. Cited on page 19.

DASTJERDI, R.; MONTAZER, M. A review on the application of inorganic nano- structured materials in the modification of textiles: Focus on anti-microbial properties. *Colloids and Surfaces B: Biointerfaces,* v. 79, n. 1, p. 5 -18, 2010. ISSN 0927-7765. Available at:

<http://www.sciencedirect.com/science/article/pii/S0927776510001773>. Cited on page 21.

DAVID, N. C. et al. Electro-conductive fabrics based on dip coating of cotton in poly(3-hexylthiophene). v. 28, n. 5, p. 583 - 589, 5 2017. Available at: <http://doi.wiley.com/10.1002/pat.3857>. Cited 3 times on pages 11, 22 and 23.

DOWLING, D. P. et al. Deposition of anti-bacterial silver coatings on polymeric substrates. In:. [S.L: s.n.], 2001. v. 398-399, p. 602 - 606. Cited on page 25.

DUBAS, S. P.; KUMLANGDUDSANA, P.; POTIYARAJ. Layer-by-layer deposition of antimicrobial silver nanoparticles on textile fibres. v. 289, n. 1-3, p. 105-109 -, 2006. Cited on page 51.

DURAN, N. E. et aL Antibacterial effect of silver nanoparticles produced by fungai process on textile fabrics and their effluent treatment. v. 3, n. 2, p. 203 - 208 2007. Available at: <http://www.lqes.iqm.unicamp.br/images/publicacoes_teses_trabalhosrev_2007_antibacteriaLpdf>. Cited 4 times on pages 48, 50, 51 and 53.

DURST, O.; ELLERMEIER, J.; BERGER, C. Influence of plasma-nitriding and surface roughness on the wear and corrosion resistance of thin films (PVD/PECVD). v. 203, n. 5, p. 848-854, 2008. Available at: <http://www.sciencedirect.com/science/article/ pii/S0257897208003514>. Cited twice on pages 11 and 28.

ECHER, F. R. et aL *Potassium sources in the top dressing of cotton yield, foliar diagnosis, fibre quality and economic analysis.* 2008. Dissertation (Master's) - Universidade do Oeste Paulista. Available at: <http: //tede.unoeste.br/tede/tde_busca/arquivo.php?codArquivo=147>. Cited on page 17.

EGBUTA, M. A.; MWANZA, M.; BABALOLA, O. O. Health Risks Associated with Expo- sure to Filamentous Fungi. *International Journal of Environmental Research and Public Health,* v. 14, n. 7, 2017. ISSN 1660-4601. Available at: <http://www.mdpi.com/1660- 4601/14/7/719>. Cited on page 11.

EGBUTA, M. A.; MWANZA, M.; BABALOLA, O. O. Health Risks Associated with Exposure to Filamentous Fungi. v. 14, n. 7, 7 2017. Available at: <http://www.ncbi.nlm.nih.gOv/pubmed/28677641http://www.pubmedcentral.nih.gov/articlerender.fcgi?artid=PMC5551157>. Cited on page 50.

EREMENKO, A. M. et al. Antibacterial and Antimycotic Activity of Cotton Fabrics, Impregnated with Silver and Binary Silver/Copper Nanoparticles. *Nanoscale Research Letters,* Springer US, v. 11, p. 28 -, 2016. ISSN 1931-7573. Available at: <https: //www.ncbi.nlm.nih.gov/pmc/articles/PMC4717125/>. Cited on page 20.

FABRICATING the Body: Textiles and human health in historical perspective. *Textile history,* v. 42, n. 2, p. 261 -, 11 2011. ISSN 0040-4969. Available at: <https://www. ncbi.nlm.nih.gov/pmc/articles/PMC4398981/>. Cited on page 16.

FERRARI, A. C.; ROBERTSON, J. Interpretation of Raman spectra of disordered and amorphous carbon. *Phys. Rev. B,* American Physical Society, v. 61, p. 14095 - 14107, May 2000. Available at: <http://link.aps.Org/doi/10.1103/PhysRevB.61.14095>. Cited on page 24.

FIJAN, S. *Microorganisms with claimed probiotic properties:* An overview of recent literature. 2014. Cited on page 20.

FIJAN, S. et al. Antimicrobial disinfection effect of a laundering procedure for hospital textiles against various indicator bacteria and fungi using different substrates for simulating human excrements.

2007. Cited on page 20.

FIJAN, S.; SOSTAR-TURK, S.; CENCIC, A. Implementing hygiene monitoring systems in hospital laundries in order to reduce microbial contamination of hospital textiles. *Journal of Hospital Infection,* Elsevier, v. 61, n. 1, p. 30 - 38, 2005. ISSN 0195-6701. Doi: 10.1016/j.jhin.2005.02.005. Cited on page 20.

GROF, G. L. et al. *Administrative practices in Uruk between 3500 and 2900 BC.* 2013. Dissertation (Master's) - University of São Paulo. Available at: <http://www. teses.usp.br/teses/disponíveis/8/8138/tde-08012014-153942/>. Cited on page 16.

HETTLICH, H. et al. Plasma-induced surface modifications on suicone intraocular lenses: Chemical analysis and in vitro characterisation. *Biomaterials,* v. 12, n. 5, p. 521 - 524, 1991. ISSN 0142-9612. Available at: <http://www.sciencedirect.com/science/ article/pii/0142961291901532>. Cited on page 28.

HOLBROOK, R. D.; RYKACZEWSKI, K.; STAYMATES, M. E. Dynamics of silver nano-particle release from wound dressings revealed via in situ nanoscale imaging. *Journal of Materials Science. Materials in Medicine,* Springer US, v. 25, n. 11, p. 2481 - 2489, 2014. ISSN 0957-4530. Available at: <https://www.ncbi.nlm.nih.gov/pmc/articles/ PMC4198808/>. Cited on page 21.

HOMAN, K. et al. Silver nanosystems for photoacoustic imaging and image-guided therapy. *Journal of Biomedical Optics,* Society of Photo-Optical Instrumentation En- gineers, v. 15, n. 2, p. 021316 -, 2010. ISSN 1083-3668. Available at: <https: //www.ncbi.nlm.nih.gov/pmc/articles/PMC2859084/>. Cited on page 21.

JENSEN, K. C. et al. Isolation and Host Range of Bacteriophage with Lytic Activity against Methicillin-Resistant *Staphylococcus aureus* and Potential Use as a Fomite Decontaminant. *PLoS ONE,* Public Library of Science, v. 10, n. 7, p. e0131714 -, 2015. ISSN 1932-6203. Available at: <https://www.ncbi.nlm.nih.gov/pmc/articles/ PMC4488860/>. Cited on page 20.

JONES, T. M.; LUTZ, E. a. Environmental survivability and surface sampling efficiencies for Pseudomonas aeruginosa on various fomites. 2014. Cited on page 19.

JUNQUEIRA, L. C.; CARNEIRO, J. *Basic histology.* 12th ed. Rio de Janeiro: Guanabara Koogan, 2013. Cited on page 16.

KELLY, R; ARNELL, R. Magnetron sputtering: a review of recent developments and applications. *VACUUM,* PERGAMON-ELSEVIER SCIENCE LTD, v. 56, n. 3, p. 159 - 172, 3 2000. ISSN 0042-207X. Meeting on Recent Advances in Surface Engineering, MANCHESTER METROPOLITAN UNIV, MANCHESTER, ENGLAND, MAR 31, 1999. Available at: <http://linkinghub.elsevier.com/retrieve/pii/S0042207X9900189X>. Cited 5 times on pages 11, 28, 40, 42 and 43.

KELLY, R; ARNELL, R. Magnetron sputtering: a review of recent developments and applications. *Vacuum,* v. 56, n. 3, p. 159 - 172, 2000. ISSN 0042-207X. Available at: <http://www.sciencedirect.com/science/article/pii/S0042207X9900189X>. Cited on page 25.

KRAMER, A.; SCHWEBKE, I.; KAMPF, G. *How long do nosocomialpathogens persist on inanimate surfaces? A systematic review.* 2006. Cited on page 19.

KUZDAN, C. et al. Three-year study of health care-associated infections in a Turkish paediatric ward. 2014. Cited twice on pages 19 and 20.

LABBÉ, E. N. et al. Evidence of fatal skeletal injuries on Malapa Hominins 1 and 2. *Scientific Reports,*

Nature Publishing Group, v. 5, p. 15120 -, 2015. ISSN 2045-2322.

Available at: <https://www.ncbi.nlm.nih.gov/pmc/articles/PMC4602312/>. Quoted on page 15.

LAPORT, M. S. et al. Heat-resistance and heat-shock response in the nosocomial pathogen Enterococcus faecium. 2003. Cited on page 20.

LEE, H.; YEO, S.; JEONG, S. Antibacterial effect of nanosized silver colloidal solution on textile fabrics. *JOURNAL OF MATERIALS SCIENCE,* KLUWER ACADEMIC PUBL, VAN GODEWIJCKSTRAAT 30, 3311 GZ DORDRECHT, NETHERLANDS, v. 38, n. 10, p. 2199 - 2204, MAY 15 2003. ISSN 0022-2461.2nd International Conference on Polymer Fibres 2002, MANCHESTER, ENGLAND, JUL 10-12, 2002. Quoted on page 53.

LEE, H. J. Antibacterial effect of nanosized silver colloidal solution on textile fabrics. v. 38, n. 10, p. 2199-2204, 2003. Available at: <http://link.springer.eom/10.1023/A: 1023736416361>. Cited 3 times on pages 48, 50 and 51.

LEITE, A. da S. et al. *Textile industry.* 2015. Dissertation (Master's) - University of São Paulo. Available at: <http://www.teses.usp.br/teses/disponiveis/100/100133/tde- 21062015-224857/>. Cited on page 16.

LI, X. et al. Conversion of a Metastable Superhydrophobic Surface to an Ultraphobic Surface. v. 24, n. 15, p. 8008 - 8012, 8 2008. Available at: <http://pubs.acs.org/doi/ abs/10.1021/la801044j>. Cited on page 48.

LIEBERMAN, A. J. L. M. A. *Principies of Plasma Discharges and Materials Processing. 2°.* ed. New York:: Wiley, 2005. ISBN 978-0-471-72001-0. Available at: <http://www. wiley.com/WileyCDA/WileyTitle/productCd-0471720011 .html>. Cited on page 26.

LIMA, R. J. C. de et al. *The effects of trade liberalisation on the Brazilian textile industry.* 2006. Dissertation (Master's) - Rio de Janeiro State University. Available at: <http://www.bdtd.uerj.br/tde_busca/arquivo.php?codArquivo=744>. Cited on page 16.

LÓPEZ-GIGOSOS, R. et al. Persistence of nosocomial bacteria on 2 biocidal fabrics based on silver under conditions of high relative humidity. 2014. Cited on page 19.

LUQUETA, G. H. et al. Wireless Sensor NetWork to Monitoring an Ozone Steriliser. *IEEE Latin America Transactions,* v. 14, n. 5, p. 2167 - 2174, 2016. Available at: <http://ieeexplore.ieee.org/document/7530410/>. Cited on page 31.

MARCIANO, F. et al. Antibacterial activity of \{DLC\} and AgâDLC films produced by \{PECVD\} technique. *Diamond and Related Materials,* v. 18, n. 5â8, p. 1010 - 1014, 2009. ISSN 0925-9635. Proceedings of Diamond 2008, the 19th Euro- pean Conference on Diamond, Diamond-Like Materials, Carbon Nanotubes, Nitrides and Silicon Carbide. Available at: <http://www.sciencedirect.com/science/article/pii/ S0925963509001058>. Cited on page 28.

MARCIANO L.F. BONETTI, L. S. N. D. E. C. V. T. F. Antibacterial activity of DLC and Ag-DLCfilms produced by PECVD technique. v. 18, p. 1010-1014 -, 2009. Cited on page 26.

MATUSSEK, A.; TAIPALENSUU, J.; EINEMO, I. Transmission of Staphylococcus aureus from maternity unit staff members to newborns disclosed through spa typing. *American Journal of Infection Control,* Elsevier, v. 35, n. 2, p. 122 - 125, 2007. ISSN 0196-6553. Doi: 10.1016/j.ajic.2006.08.009. Cited on page 20.

MEJÍA, M. I. et al. Magnetron-Sputtered Ag Surfaces. New Evidence for the Nature of the Ag Ions

Intervening in Bacterial Inactivation. *ACS Applied Materials & Interfaces,* American Chemical Society, v. 2, n. 1, p. 230 - 235, 1 2010. ISSN 1944- 8244. doi: 10.1021/am900662q. Available at: <http://pubs.acs.org/doi/abs/10.1021/ am900662q>. Cited 6 times on pages 37, 40, 42, 43, 51 and 55.

MINEA, T. M. et al. Single chamber PVD/PECVD process for in situ control of the catalyst activity on carbon nanotubes growth. v. 200, n. 1, p. 1101 - 1105, 2005. Available at: <http://www.sciencedirect.com/science/article/pii/S0257897205001271>. Cited twice on pages 12 and 28.

MITCHELL, A.; SPENCER, M.; EDMISTON, C. *Role of healthcare apparel and other healthcare textiles in the transmission of pathogens* A review of the literature. 2015. Cited on page 19.

MITRA, C. et al. Citrate-Coated Silver Nanoparticles Growth-Independently Inhibit Aflatoxin Synthesis in Aspergillus parasiticus. *Environmental Science & Technology,* ACS publications, v. 1, n. 9, p. 1 - 38, jun 2017. PMID: 28618218. Available at: <http://pubs.acs.Org/doi/abs/10.1021/acs.est.7b01230>. Cited on page 11.

MITRA, C. et al. Citrate-Coated Silver Nanoparticles Growth-Independently Inhibit Aflatoxin Synthesis in Aspergillus parasiticus. v. 51, n. 14, p. 8085 - 8093, 7 2017. Available at: <http://www.ncbi.nlm.nih.gov/pubmed/28618218>. Cited on page 50.

MIYAZAKI, N. H. T; BÔAS, M. H. S. V. *Molecular analysis associated with the study of resistance genes in methicillin-resistant Staphylococcus aureus.* 2006. Thesis (Doctorate). Available at: <http://157.86.8.70:8080/certifica/icict/3101>. Cited on page 19.

MORAES, A. C. M. de et al. *Graphene oxide and graphene oxide functionalized with silver nanoparticles.* 2015. Thesis (PhD) - State University of Campinas. Institute of Chemistry. Available at: <http://libdigi.unicamp.br/document/?code= 000951712>. Cited on page 21.

MOURA, S. *Química Geral,* sexta. Porto Alegre: Artmed, 2007. Quoted on page 24.

OHRING, M. *Materials Science of Thin Films.* First. San Diego: Academic Press, 1992. Cited on page 41.

OHRING, M. Chapter 4 - Discharges, Plasmas, and Ion-Surface Interactions. *Materials Science of Thin Films,* p. 145 - 202, 2002. Cited on page 41.

OLIVEIRA, A. C. X. de et al. *From Uruk to the Hadrianic Villa.* 2007. Thesis (Doctorate) - University of São Paulo. Available at: <http://www.teses.usp.br/teses/disponiveis/ 16/16131/tde-19112010-144043/>. Cited on page 16.

PANAGEA, S. et al. Environmental contamination with an epidemic strain of <em>Pseudomonas aeruginosa</em> in a Liverpool cystic fibrosis centre, and study of its survival on dry surfaces. *Journal of Hospital Infection,* Elsevier, v. 59, n. 2, p. 102 - 107, 2005. ISSN 0195-6701. Doi: 10.1016/j.jhin.2004.09.018. Cited on page 19.

PASINATO, T. L. S. et al. *The textile production system in Brazil.* 2016. Master's dissertation. Available at: <http://hdl.handle.net/11338/1418>. Cited twice on pages 17 and 18.

PERELSHTEIN, I. et al. Sonochemical coating of silver nanoparticles on textile fabrics (nylon, polyester and cotton) and their antibacterial activity. v. 19, n. 24, p. 245705 -, 6 2008. Disponível em: <http://stacks.iop.org/0957-4484/19/i=24/a=245705?key=crossref. 30449742d23eacaba16912ff86dccaca>. Cited twice on pages 50 and 53.

PESSOA, R. S. *Studies of fluorinated plasmas applied to silicon corrosion using global model simulation and experimental diagnostics.* 2009. Thesis (Doctorate) - Technological Institute of Aeronautics, São José dos Campos. Cited twice on pages 26 and 27.

PFENNIG, D. W.; PFENNIG, K. S. Character displacement and the origins of diversity. *The American Naturalist,* v. 176, n. Suppl 1, p. S26 - S44,12 2010. ISSN 0003-0147. Available at: <https://www.ncbi.nlm.nih.gov/pmc/articles/PMC3285564/>. Cited on page 16.

R, M. A. R. et al. An evaluation of the tribological characteristics of DLC films grown on Inconel Alloy 718 using the Active Screen Plasma technique in a Pulsed-DC PECVD system. *Surface and Coatings Technology,* v. 284, p. 235 - 239,12 2015. ISSN 0257- 8972. Available at: <http://linkinghub.elsevier.com/retrieve/pii/S0257897215005411>. Cited 3 times on pages 11, 28 and 30.

RABUZA, U.; SOSTAR-TURK, S.; FIJAN, S. Efficiency of four sampling methods used to detect two common nosocomial pathogens on textiles. *TEXTILE RESEARCH JOURNAL,* v. 82, n. 20, p. 2099 - 2105, DEC 2012. ISSN 0040-5175. Cited on page 11.

RADETIC, M. Functionalisation of textile materials with silver nanoparticles. *JOURNAL OF MATERIALS SCIENCE,* SPRINGER, 233 SPRING ST, NEW YORK, NY 10013 USA, v. 48, n. 1, p. 95 - 107, JAN 2013. ISSN 0022-2461. Cited 4 times on pages 11, 16, 19 and 22.

RADETIc, M. Functionalisation of textile materials with silver nanoparticles. *Journal of Materials Science,* v. 48, n. 1, p. 95 - 107, 1 2013. ISSN 1573-4803. Available at: <https://doi.org/10.1007/s10853-012-6677-7>. Cited on page 51.

REDFORD, D. B. *Egypt, Canaan and Israel in Ancient Times.* Princeton: Princeton University Press, 1993. Quoted on page 16.

RI VERO, P. J. et al. An antibacterial coating based on a polymer/sol-gel hybrid matrix loaded with silver nanoparticles. *Nanoscale Research Letters,* Springer, v. 6, n. 1, p. 305 2011. ISSN 1931-7573. Available at: <https://www.ncbi.nlm.nih.gov/pmc/articles/ PMC3211391 />. Cited on page 25.

ROBERTSON, J. Diamond-like amorphous carbon. *Materials Science and Engineering: R: Reports,* v. 37, n. 4-6, p. 129 - 281, 5 2002. ISSN 0927-796X. Available at: <http://linkinghub.elsevier.com/retrieve/pii/S0927796X02000050>. Cited 3 times on pages 25, 26 and 28.

SABIA, C. et al. Detection of bacteriocin production and virulence traits in vancomycin- resistant enterococci of different sources. *JOURNAL OFAPPLIED MICROBIOLOGY,* v. 104, n. 4, p. 970 - 979, APR 2008. ISSN 1364-5072. Cited on page 20.

SAMBERG, M. E.; ORNDORFF, P. E.; MONTEIRO-RIVIERE, N. A. Antibacterial efficacy of silver nanoparticles of different sizes, surface conditions and synthesis methods. v. 5, n. 2, p. 244 - 253, 2011. Cited on page 20.

SANTOS, E. et al. Macrophages adhesion rate on TÍ-6AI-4V substrates: polishing and DLC coating effects. *Research on Biomedical Engineering,* scielo, v. 32, n. 2, p. 144 - 152, 6 2016. ISSN 2446-4740. Available at: <http://www.scielo.br/scielo.php?script= sci_arttext&pid=S2446-47402016000200144&lng=en&tlng=en>. Cited on page 52.

SATAEV, M. et al. Novel process for coating textile materials with silver to prepare antimicrobial fabries. *Colloids and Surfaces A: Physicochemical and Engineering As-* pects, v. 442, p. 146 - 151,2

2014. ISSN 0927-7757. Selected papers from the 26th European Colloid and Interface Society conference (26th {ECIS} 2012). Available at: <http://linkinghub.elsevier.com/retrieve/pii/S0927775713001210>. Cited on page 51.

SHARMA, V. K.; YNGARD, R. A.; LIN, Y. Silver nanoparticles: Green synthesis and their antimicrobial activities. *ADVANCES IN COLLOID and INTERFACE SCIENCE*, ELSEVIER SCIENCE BV, v. 145, n. 1, p. 83 - 96, 2009. ISSN 0001 -8686. Available at: <http://www.sciencedirect.com/science/article/pii/S0001868608001449>. Cited on page 20.

SHAW, I. The Oxford History of Ancient Egypt. In:. New York: Oxford University Press, 2002. ch. 10. quoted on page 16.

SILEIKAITE, A. et al. Analysis of Silver Nanoparticles Produced by Chemical Reduction of Silver Salt Solution. v. 12, n. 4, p. 287 - 291,2006. Cited twice on pages 23 and 25.

SISSON, G. et al. Randomised clinical trial: A liquid multi-strain probiotic vs. Placebo in the irritable bowel syndrome - A 12 week double-blind study. 2014. Cited on page 20.

SOARES, B. F.; MACIEL, A. J. da S. *Operational performance of an electrostatic spray tip for phytosanitary treatment of the cotton crop (Gossypium sp).* 2001. Dissertation (Master's Degree) - State University of Campinas. Faculty of Agricultural Engineering. Available at: <http://libdigi.unicamp.br/document/ ?code=vtls000287894>. Cited on page 17.

SOUZA, G. D. de et al. Silver: Brief history, properties and applications. *Educación química,* scielomx, v. 24, p. 14 - 16, 01 2013. ISSN 0187-893X. Available at: <http://www.scielo.org.mx/scielo.php?script=sci_arttext&pid=S0187-893X2013000100003&nrm=iso>. Cited on page 21.

STRINGHAM, S. A. et al. Divergence, convergence, and the ancestry of feral populations in the domestic rock pigeon. *Current Biology,* v. 22, n. 4, p. 302 - 308, 2 2012. ISSN 0960-9822. Available at: <https://www.ncbi.nlm.nih.gov/pmc/articles/PMC3288640/>. Cited on page 16.

SUN, N.; WANG, T.; YAN, X. Self-assembled supermolecular hydrogel based on hydroxyethyl cellulose: Formation, in vitro release and bacteriostasis application. v. 172, p. 49 - 59, 9 2017. Available at: <http://linkinghub.elsevier.com/retrieve/ pii/S0144861717305337>. Cited on page 53.

SZOSTAK-KOTOWA, J. Biodeterioration of textiles. *INTERNATIONAL BIODETERIO- RATION & BIODEGRADATION,* v. 53, n. 3, p. 165 - 170, 4 2004. ISSN 0964-8305. 2nd Polish Microbial Biodeterioration Symposium of Technical Materials, Lodz, PO- LAND, MAY 30-31, 2001. Available at: <http://linkinghub.elsevier.com/retrieve/pii/ S0964830503000908>. Quoted on page 11.

TANG, X. et al. Facile dip-coating process towards multifunctional nonwovens: Robust noise reduction, abrasion resistance and antistatic electricity. p. 004051751772512 -, 8 2017. Disponível em: <http://journals.sagepub.com/doi/10.1177/0040517517725120>. Cited twice on pages 11 and 22.

THOMSON REUTERS. *Report of searches for papers related to the topic "Silver nanoparticles".* 2018. Available at: <https://apps.webofknowledge.com/Search. do?product=WOS&SID=5D6SoxMnG6v1sLZThJT&search_mode=GeneralSearch& prID=7b581318-f89e-4471 -afe3-06ef3b5e967b>. Accessed on: 09/04/2018. Cited on page 21.

VENKATESH, M.; TAKTAK, S.; MELETIS, E. I. Synthesis of Ag-doped hydrogenated carbon thin films by a hybrid PVD-PECVD deposition process. v. 37, n. 7, p. 1669 - 1676, 2014. Available at: <https://doi.org/10.1007/s12034-014-0728-4>. Cited twice on pages 12 and 28.

VU, N. K. et al. Effect of Particle Size on Silver Nanoparticle Deposition onto Dielectric Barrier Discharge (DBD) Plasma Functionalised Polyamide Fabric. *Plasma Processes and Polymers,* WILEY-VCH Verlag, v. 10, n. 3, p. 285 - 296, 2013. ISSN 1612-8869. Available at: <http://dx.doi.org/10.1002/ppap.201200089>. Cited twice on pages 48 and 52.

WANG, H. et al. Preparation and characterisation of silver nanocomposite textile. *JOURNAL OF COATINGS TECHNOLOGY and RESEARCH,* SPRINGER, v. 4, n. 1, p. 101 - 106, 4 2007. ISSN 1547-0091. Available at: <http://link.springer.com/10.1007/s11998- 007-9001-8>. Cited 3 times on pages 37, 51 and 55.

WANG, H. et al. Preparation and characterisation of silver nanocomposite textile. *Journal of Coatings Technology and Research,* v. 4, n. 1, p. 101 - 106, Mar 2007. ISSN 1935- 3804. Cited twice on pages 51 and 55.

WILD, C. et al. Proceedings of the 4th European Conference on Diamond, Diamond- like and Related Materials Oriented CVD diamond films: twin formation, structure and morphology. *Diamond and Related Materials,* v. 3, n. 4, p. 373 - 381, 1994. ISSN 0925-9635. Available at: <http://www.sciencedirect.com/science/article/pii/ 0925963594901880>. Cited on page 28.

ZHANG, J. et al. Facile preparation of durable and robust superhydrophobic textiles by dip coating in nanocomposite solution of organosilanes. v. 49, n. 98, p. 11509 -, 2013. Available from: <http://xlink.rsc.org/?DOI=c3cc43238f>. Cited 4 times on pages 11, 22, 23 and 28.

ZHANG, J. et al. Phytoliths reveal the earliest fine reedy textile in China at the Tianlu- oshan site. *Scientific Reports,* Nature Publishing Group, v. 6, p. 18664-, 2016. ISSN 2045-2322. Available at: <https://www.ncbi.nlm.nih.gov/pmc/articles/PMC4725870/>. Cited on page 16.

ZILLE, A. et al. Size and Aging Effects on Antimicrobial Efficiency of Silver Nanoparticles Coated on Polyamide Fabrics Activated by Atmospheric DBD Plasma. *ACS Applied Materials & Interfaces,* v. 7, n. 25, p. 13731 -13744, 2015. PMID: 26057400. Available at: <http://dx.doi.Org/10.1021/acsami.5b04340>. Cited 4 times on pages 20, 34, 48 and 52.

ZEMLIcKA, R. et al. Principies and practice of an automatic process control for the deposition of hard nc-TiC/a-C:H coatings by hybrid PVD-PECVD under industrial conditions. v. 304, n. Supplement C, p. 9 - 15, 2016. Available at: <http: //www.sciencedirect.com/science/article/pii/S0257897216305539>. Cited twice on pages 12 and 28.

Printed by Books on Demand GmbH, Norderstedt / Germany